POLLUTION CONTROL TECHNOLOGIES FOR SMALL-SCALE OPERATIONS

DECEMBER 2023

ASIAN DEVELOPMENT BANK

Contents

Appendixes

Tables, Figures, and Box

Tables

Figures

Box

Acknowledgments

The study was initiated and led by Xin Ren, senior environmental specialist, Office of Safeguards of the Asian Development Bank (ADB), funded by the regional technical assistance (TA 9647) on Strengthening Safeguards Management in Southeast Asia. Xin Ren authored Chapters 1, 3, 4, and 6 and performed quality control for the entire report. Chirong Huang, senior consultant and sanitation engineer of more than 30 years of experience, drafted the rest of the report and contributed to Chapter 4.

Xin Ren has worked closely in rural projects of Southeast Asian countries, focusing on environmental management. Before joining ADB, she worked in the World Bank in diverse projects ranging from urban, energy, and transport to rural sectors. Apart from her 15 years in multilateral development banks, she also worked in United Nations agencies on climate change and waste management for years, and in the People's Republic of China on pollution control, hazardous waste management, and cleaner production. This study draws upon the two authors' substantial experience in different facets of pollution control and their insights on broader environmental management in developing countries.

The study report benefited from the valuable comments and suggestions by peer reviewers, James G. Baker, senior circular economy specialist, ADB; and Jitendra K. Singh, water supply and sanitation specialist, ADB. The authors also want to thank Antoine Morel, coordinator of environmental work in Southeast Asia Department of ADB and TA 9647, for his support all through, and Ricardo Carlos Barba, officer in charge of Regional Office 2 in the Office of Safeguards of ADB, for his guidance in review and publication. The authors remain responsible for all the residual shortcomings.

Abbreviations

BAT	best available technology
BOD5	five-day biochemical oxygen demand
BORDA	Bremen Overseas Research and Development Association
BPT	best practicable currently available technology
CAPEX	capital expenditure
COD	chemical oxygen demand
DEWATS	decentralized wastewater treatment system
DMC	developing member country
EHS	environmental, health, and safety
EIA	environmental impact assessment
GAC	granular activated carbon
GHG	greenhouse gas
HDT	hydraulic detention time
IFC	International Financial Corporation
MDB	multilateral development bank
MLSS	mixed liquor suspended solids
O&G	oil and grease
O&M	operation and maintenance
OPEX	operating expense
PRC	People's Republic of China
RAS	returned active sludge
RBC	rotating biological contactor
SMEs	small and medium-sized enterprises
SRT	sludge retention time
TSS	total suspended solids
UK	United Kingdom
US	United States
US EPA	US Environmental Protection Agency
WAS	waste activated sludge
WSP	waste stabilization pond
WWTP	wastewater treatment plant

Executive Summary

Agriculture activities are very diverse and deal with vastly different raw materials, products, and production technologies. Many of them, especially those related to animals, generate pollution that is equally diverse and complex. In particular, wastewater and solid waste from animal husbandry and processing have become one of the largest polluters in many rural areas. With the coronavirus disease (COVID-19) pandemic and heightened awareness, pathogenic and infectious risks of agro-sectors also received more attention.

The Asian Development Bank has supported agriculture and rural development for decades. Most of these activities are small scale in rural areas with very limited resources and capacity. This adds more challenges in controlling pollution and complying with environmental, health, and safety (EHS) requirements. It is therefore critical to identify, evaluate or validate, and select the suitable methods and technologies that are effective on EHS yet affordable, easy to operate, and thus sustainable.

Yet the project developers/technical designers for these production activities are not experts on pollution control, let alone its technologies. Most practitioners for environmental impact assessment (EIA) are not engineers with the expertise to enable them to perform validation or selection. There is a lack of systematic comparative study of the cost-effectiveness of major technologies suitable for small-scale operations, whether in rural or urban settings, for people to readily tap into to meet the need of projects in general and in the feasibility study and EIA in particular.

In this study, major control methods and treatment technologies for wastewater, organic solid and bio-hazardous wastes, chemical substances, and air emission typical for the agro-sector are reviewed to evaluate their suitability for small-scale operations. It focuses on their efficiency and cost-effectiveness in reducing main pollutants including microbes regulated. It includes investment and operation and maintenance cost in relative terms, in addition to efficiency and main pros and cons for technologies covered and compared.

The most salient indicator for technologies' ability to abate pollution is their removal efficiency of targeted pollutants. However, the discussion shows that specific design for the pollution at hand and operational factors can greatly affect the results of pollution treatment or waste utilization. Therefore, the efficiencies of major technologies serve mostly as a starting point and reference. With them, it is hoped that project designers and EIA preparers can at least judge the validity and feasibility of various technologies proposed to them in controlling the pollution and EHS risks in question.

The discussion on wastewater also illustrates the critical role played by environmental standards, to which a chapter is dedicated. It describes the difference between two main types of standards, ambient versus discharge, in terms of their basis and implications. Wastewater standards of most developing countries in Southeast Asia and in the People's Republic of China are compared with that of representative developed countries as well as their approaches to gain insights and provide recommendations.

Given that cost-effectiveness is the key for small-scale pollution control, a repeated theme regardless of the type of pollution is segregation, i.e., separating liquid from solids, cleaner from dirtier stream, or organic from hazardous, in order to control, utilize, and treat cost-effectively. The results can be useful not only for the agro-sector but also for the peri-urban, industries, or health sectors that need to deal with small-scale yet complex pollution, waste management, and biosafety issues.

Different technologies work best for certain types of pollution in a certain range of strength. A lack of decent knowledge on pollution characteristics will result in improper choice of methods, technologies, or their designed capacity, leading to ineffective or unsustainable pollution abatement or even failure. Therefore, a study on pollution characteristics for typical agro-sectors is being undertaken in parallel and its results are to be used in tandem with this report.

1 Introduction

Background

Agriculture activities are very diverse, mainly including horticulture and plant-based processing, animal husbandry and processing, aquaculture and fish processing, laboratories, manufacturing of vaccines and other veterinary products, agro-waste utilization, and by-products production. They deal with vastly different raw materials, products, and production technologies. Many of them, especially those related to animals, generate pollution that is equally diverse and complex with impacts and risks to water, air, soil, ecosystem, and living resources along the food chain. Notably, wastewater and solid wastes from animal husbandry and processing have become one of the largest polluters in many rural areas. In addition, they involve activities with pathogenic and infectious risks, brought to the front by the coronavirus disease (COVID-19) pandemic.

To tackle the environmental, health, and safety (EHS) issues of agro-sectors, it is most important to understand their pollution characteristics, namely the amount of pollution, major pollutants and their concentration range, or their nature (e.g., organic or inorganic and types of hazards). These lay the foundation for the next step, to control and abate the pollution to meet applicable standards and requirements.

Many rural projects of the Asian Development Bank (ADB) support cooperatives or small and medium-sized enterprises (SMEs) in the abovementioned agro-sectors. During preparation of these projects, lack of expertise and time during the feasibility study and environmental impact assessment (EIA) often hinder the evaluation, validation, or selection of suitable pollution control methods and technologies that can fulfill compliance requirements yet be affordable and sustainable. Technologies that are either active and/or promoted in the country by private companies or bilateral development agencies often prevail.

Therefore, funding was obtained from Technical Assistance (TA) (TA 9647: Strengthening Safeguards Management in Southeast Asia) to carry out two parallel studies to fill these gaps and address the difficulties facing similar ADB projects, in particular in the environment and rural sector. Nonetheless, results of this study on pollution control technologies can also be useful for projects in peri-urban, energy efficiency in industry, or health sectors that need to deal with small-scale wastewater treatment and solid waste management.

Objectives

To fill the abovementioned gaps, major control and treatment technologies for wastewater, solid waste, and air pollution for agro-sectors are reviewed to evaluate or verify their suitability for small-scale and even medium-scale operations. The evaluation focuses on their efficiency in reducing major pollutants regulated in applicable environmental standards, e.g., suspended solids, biochemical oxygen demand (BOD), and bacterial parameters. These elements are needed for the EIA process to evaluate possibility of compliance with applicable standards after pollution treatment by different technical options recommended by the project (alternative comparison) and development of environmental management plans.

The study also looks at other key factors such as investment capital expenditure (CAPEX), ease in operation and maintenance (O&M), and cost. These are not only indispensable for the feasibility study and costing of such projects, but also useful for the selection and technical design of their pollution control. The work can benefit other types of rural projects and also those in a peri-urban setting.

Prior to selecting among available treatment technologies for the pollution in question, its characteristics and reduction requirements need to be thoroughly investigated. Without full and clear understanding of the pollution characteristics, a pollution treatment facility is doomed as is apparent in many failures. Such investigation is completed in another study on pollution characteristics funded under TA-9647.

However, every pollution management facility has its own unique characteristics in terms of quantity and quality of influent wastewater and solid waste. Not only water and material usage pattern but also collection efficiency will vary among projects. To determine the pollution characteristics to be treated, the reports of both TA studies cannot substitute for the investigation and tests specific for the pollution at hand.

Methodology and Coverage

Geographic scope of work is intended for but not limited to developing member countries (DMCs) of ADB in Southeast Asia. This means the pollution control technologies studied should be applicable to tropical and subtropical zones.

The main methodologies employed are as follows:

(i)   Literature survey and review: including ADB project documents and data;
(ii)   Research of relevant discharge standards and limits: relying on unofficial English translation of national environmental standards and clarification by national environmental staff and consultants; and
(iii)   Relative cost level estimated to the extent possible: based on the authors' experience, expert opinions emerged from various projects, in addition to the literature survey.

Principles

Separation of solid waste from wastewater is needed for greater efficiency in wastewater treatment and reuse so as to minimize unnecessary treatment requirements and cost. Removing any solid waste from liquid waste will benefit downstream treatment units and reduce the equipment needed—thus also CAPEX, related operating expense (OPEX), and O&M burden.

Meanwhile, excessive water usage from routine production and processing may generate significant and unnecessary liquid waste, which could affect the treatment volume and efficiencies. Thus, water usage for washing, flushing, and/or processing all need to be well controlled to avoid extra CAPEX and OPEX. Overall, wastewater minimization and reuse should be promoted as long as permitted by hygiene and food safety requirements. This not only saves valuable water but also produces less diluted wastewater, which, in turn, generally requires less energy cost per unit of pollution reduced. Separation of solids from wastewater and segregation of different wastewater streams to the extent possible ensures a more cost-effective and sustainable pollution management.

Typical wastewater treatment consists of three levels: primary (physical), secondary (biological), and tertiary (advanced) treatment processes. Conventionally, the first two levels of wastewater treatment are sufficient for reasonable required discharge limits. Unless more stringent requirements for wastewater treatment plant (WWTP) effluent are set, such as on nitrogen and phosphorus removal, tertiary treatment is not normally required.

Primary treatment generally utilizes physical–chemical methods to remove large solids, plastic materials, rags, rocks, and sand from the influent wastewater. Typical primary treatment or system equipment includes bar screens (coarse and fine, mechanically or fixed), grit chambers, primary sedimentation tanks, and sometimes a flow equalization basin for quantity and quality balance. The primary sedimentation tank is the key unit and is generally equipped with a sludge scraper to collect settled solids into the sludge hopper with a sludge pump to transport the naturally thickened sludge to the sludge treatment units. This removes about 40%–50% of floatable solids and total suspended solids (TSS) and about 30%–40% of BOD based on many WWTP operational records.

Secondary treatment primarily uses the biological process to decompose organic compounds. It can be a combination of aerobic and anaerobic treatment processes to provide better reduction of pollutant concentrations and higher removal rates of organic compounds in the wastewater. Effluent of the biological aeration tanks flows into secondary sedimentation tanks where concentrated solids settle toward the bottom and clear effluent flows out for further treatment, which is typically disinfection or tertiary treatment if necessary, before discharging into the receiving environment.

The settled and concentrated solids are collected and transmitted back to the biological aeration tanks as returned activated sludge (RAS), with excessive sludge being removed. Generally, effluent from the primary sedimentation tanks flows into aeration tanks to be mixed with RAS to maintain well-balanced mixed liquor suspended solids (MLSS) concentrations (typically 3,000–4,000 mg/L) according to the wastewater characteristics and applicable discharge limits. Operators normally adjust the rates of RAS and effluent to accommodate the loading variations in both quantity and quality. The standard secondary treatment process can remove around 90% of BOD and 95% of TSS.

Tertiary treatment refers to further pollutant removal, primarily BOD, TSS, total nitrogen, ammonia (NH_3)-nitrogen, and total phosphorus, to meet stricter discharge standards. This mainly includes reverse osmosis, high-rate ballast sedimentation, and a de-nitrification deep-bed filter. Reverse osmosis utilizes membrane technology to remove pollutants but has high energy demand. High-rate ballast sedimentation typically utilizes chemicals and magnetite to effectively settle TSS and total phosphorus with less tankage. A de-nitrification deep-bed filter can remove total nitrogen, with adequate microorganism growth within the filter bed. This is a modified sand filter technology with chemicals added to enhance removal of TSS and total phosphorus.

Pre-treatments and Their Comparison

In accordance with the principles described earlier, a critical step is ensuring all wastes can be collected as early as possible, once they are generated on-site. Solid wastes, especially those to be mixed with liquid due to washing, flushing, cleaning, and/or processing, should be scraped or separated. This will minimize the potential foul air and odor issue as well as reduce unnecessary loading to the downstream treatment units. The collection mechanism is another important element for timely separation of liquid and solid wastes and for ensuring all wastes are well managed.

Typically, pre-treatment for industrial wastewater is strongly recommended in order to achieve the removal of unnecessary loading to the major treatment processes downstream. Such activity generally utilizes either single or multiple units based on the wastewater characteristics. Various pre-treatment units such as a scraper for solid waste removal, oil trap/grease separator for oil and grease (O&G) removal, bar screens and grit removal for larger floatable solids and sand/grit, flow equalization for liquid waste balance and adjustment, and isolated hopper for foul air and odor collection are available for different purposes.

All investments for these pre-treatment units and equipment are not high but can significantly reduce the CAPEX and OPEX of the subsequent treatment, particularly regarding energy consumption. In fact, contributions of pre-treatment units, either single or in combination, are critical for overall wastewater treatment performance and efficiency.

Scraper

When solid wastes are generated from production and processing lines, such as manure or waste straw or other bedding materials from livestock sector operations, it is best to separate them from the liquid as early as possible. This scraping reduces the quantity of solids flowing into the WWTP and also lowers the loading to the downstream treatment.

Oil Trap/Grease Separator

Typically, O&G removal is needed for the wastewater stream from animal processing and other O&G-generating processes such as canning or fish meal production. Otherwise, these wastes not only clog collection pipelines but also undermine the downstream biological treatment process. They potentially impact the operation as well as increase CAPEX and OPEX. Thus, oil traps and grease separators are used widely and very cost-effective.

Bar Screens and Grit Removal

After most solid wastes are scraped and/or separated from the wastewater, certain floatable solids, sand, or grit flow into the next treatment process. They are normally removed by bar screens (mechanical) and grit chambers. Bar screens normally provide very limited space (3–10 millimeters) to block large and floatable material (rags, timber, and plastic bags) out of influent wastewater. A grit chamber can be aerated if necessary to increase its capability to remove silt, sand, and grit after the bar screens. All these large solids, sand, grit, shells, and scales are taken out of the treatment system to protect the downstream processes and significantly reduce energy requirements.

Flow Equalization

Generally, the amount and strength of wastewater from many activities fluctuate with the production cycle. Yet the operation of WWTP requires a much more stable flow pattern so that it will not affect the inflow pumping and the energy requirements for the key treatment process. Unstable wastewater loading will also make system operation too complicated, harder for small-scale operation, and undermine the treatment result.

Flow equalization is thus needed to even the inflows and settle part of the solids in the tank. Moreover, this additional hydraulic detention time (HDT) provides more treatment for major pollutants such as chemical oxygen demand (COD) and TSS owing to hydrolysis reaction during the flow equalization process. In addition, removal of settled solids and sand/grit from the flow equalization tank can ease the pre-treatment process and create a win–win situation for WWTP operators.

Primary Treatments and Their Comparison

Typically, the primary treatment process primarily employs physical and chemical methods to remove pollutants. The primary sedimentation tank is normally designed with adequate HDT to allow floatable solids and TSS to settle at the bottom. Settled solids (also called primary sludge) are periodically pumped and transported to sludge management units for further treatment.

Septic Tank

A septic tank is the most common primary treatment used for households, small communities, and/or SMEs as the basic facility for liquid waste. It normally provides simple but effective treatment to remove large floatable solids and untreatable wastes from the liquid waste via a bar screen and first chamber. After passing through three-stage chambers to remove some BOD, etc., the effluent can generally be reused for agricultural purposes if the HDT is adequate. Septic tanks can also serve to adjust the wastewater quantity and thicken concentration for the downstream treatment processes.

Lagoon

A lagoon typically is a series of one or two pond-like structures, originally in coastal areas using a lagoon (hence the name), to treat wastewater via natural phenomena and conditions without any energy consumption. It can be used as a primary treatment unit and is normally lined with clay or a synthetic liner to prevent/minimize wastewater seepage into the soil, groundwater, and/or adjacent waters. Lagoons utilize physical and biological processes to treat wastewater during its storage period before discharging to receiving waters or reuse for crops, pasture, and other types of land. Most treatment in a lagoon system occurs naturally by anaerobic or aerobic bacteria, depending on the design. Overall, it requires a larger land area than most treatment technologies and is more subject to weather.

In general, when discharge limits are not too stringent, these conventional primary treatments are suitable for small-scale operations, especially in developing countries. In addition, enhanced primary treatment by adding chemicals would be sufficient or as extra treatment to meet stricter discharge limits (Fang et al. 2004). This usually only needs a chemical feed system to enhance precipitation and settlement by flocculation and/or chelation. Typical chemicals used are ferric chloride, ferric sulfate, or poly-aluminum chloride. They can increase removal of BOD, TSS, and sometimes a portion of total phosphorus—all key parameters regulated in all countries (Vanotti et al. 2008). Thus, it is a good candidate as an interim treatment.

Table 2.1 summarizes actual experiences over decades, captured in the design manuals (Metcalf and Eddy 2004) issued by the largest environmental engineer association in the United States (US). *Ten States Standards* have been referenced by most design engineers for decades. These major and commonly used methods are necessary as pre-treatment or primary treatment before secondary treatment. Depending on wastewater characteristics, these simple, passive (e.g., without the need for energy to drive aeration), and thus inexpensive technologies might suffice to meet discharge standards for many small-scale operations.

Table 2.1: Primary Treatment Technologies for Wastewater

	Septic tank	Lagoon	Chemically enhanced
COD removal (%)	15–20	15–20	30–40
BOD removal (%)	20–30	30–40	40–50
TSS removal (%)	30–40	30–40	50–60
TP (%)	5–10	10–15	30–40
CAPEX relative range	Low	Low	Medium
OPEX relative range	Low	Low	Medium
Complexity in O&M	Low/medium	Low	Medium
Applicability to DMCs	High	Low	Medium

BOD = biochemical oxygen demand, CAPEX = capital expenditure, COD = chemical oxygen demand, DMC = developing member country, O&M = operation and maintenance, OPEX = operating expense, TP = total phosphorus, TSS = total suspended solids.
Source: Metcalf and Eddy (2004).

Major Technologies for Secondary Treatment

Conventional technologies for secondary/biological treatment primarily use the activated sludge process owing to its efficiency to meet discharge limits. With discharge limits being tightened over years in many countries, additional treatment processes were introduced, including the anaerobic process, thus more aggressive adjustments of RAS, etc., to more effectively remove BOD/COD, total nitrogen, and total phosphorus.

Although the activated sludge process is still the most utilized treatment technology, different variations and/or combinations have been developed that are more suited for small-scale and decentralized on-site treatment. These widely used technologies are trickling filter, rotating biological contactor (RBC), constructed wetland, aerated lagoon, and oxidation pond (also named waste stabilization pond, WSP).

In addition, the decentralized wastewater treatment system (DEWATS) by Bremen Overseas Research and Development Association (BORDA), Germany, and mini-activated sludge (called Johkasou in Japan or biotank in some countries) developed in Japan have been promoted by them in developing countries, particularly in Southeast Asia. Despite many cases of these emerging technologies, their overall treatment efficiencies and associated O&M still need more time to prove. Proven technology is especially important for SMEs in developing countries, as they generally cannot afford trial and error.

Over the years, and based on activated sludge technology, these treatment variations have been developed for better cost-effectiveness. Both trickling filter and RBC are basically designed to increase the contact between the microbes in activated sludge and wastewater to enhance decomposition without using aeration (see details in Appendix 1). Thus, they demand less energy than conventional activated sludge technology while maintaining its treatment efficacy. Other methods such as constructed wetland and oxidation pond/WSP utilize oxygen in water released from plants, algae, or by other natural forces to foster natural biodegradation processes. As a result, they require larger areas and ponds for longer retention/decomposition time instead of artificial aeration.

The difference between primary treatment and secondary treatment is not clear-cut. Notably, lagoons and septic tanks involve both physical processes to settle and biological processes to decompose pollutants. The activated sludge process is the basis of secondary treatments especially for centralized and municipal WWTP. However, it is not recommended for small-scale operations since its CAPEX and OPEX are beyond their reach (thus not covered in Table 2.2 or Appendix 1.) Some of the processes described above are variants of activated sludge, often in compacted form, e.g., the Johkasou technology developed by Japan is, in effect, a mini-activated sludge.

Table 2.2: Secondary Treatment Technologies Applicable to SMEs

	TF	RBC	CW	DEWATS	Mini-AS	AL	OP/WSP
COD removal (%)	94	80–90	60–90	85–90	80–90	65–80	31
BOD removal (%)			80–91	87–90		61–82	n/a
TSS removal (%)			80–96	93–96		40–67	
NH$_3$-N(%)	96	70–99			70–95		30
TN (%)	93	40–60	95		40–90		36
TP (%)	92		65–70		90		44
CAPEX	Medium	High		Medium	Medium/low	Medium/low	Low
OPEX	Medium	High	Low	Low	Medium	Medium	Low
O&M complexity	Medium	High	Low	Low	Medium	Medium	Low
Applicability to tropical climates	Medium	High/ medium	Medium/low	Medium/low	High/ medium	Medium/low	Low
Info sources	Zhang et al. (2015)	Waqas et al. (2023)	Various sources	BORDA 1	JSC 2	Bachi et al. (2022), etc.	Ren (2022)

AL = aerated lagoon, AS = activated sludge, BOD = biochemical oxygen demand, CAPEX = capital expenditure, COD = chemical oxygen demand, CW = constructed wetland, DEWATS = decentralized wastewater treatment system, NH3 = ammonia, O&M = operation and maintenance, OP = oxidation pond, OPEX = operating expense, RBC = rotating biological contactor, SMEs = small and medium-sized enterprises, TF = trickling filter, TSS = total suspended solids, WSP = waste stabilization pond.

Notes:

1. Given limited performance data, DEWATS efficiency can refer to that of CW due to their similarity.

2. According to the Japan Sanitation Consortium, removal rates of mini-AS are close to those of TF and RBC.

For small-scale operations typical in agro-sectors and rural or peri-urban sanitation, the cheaper and easy to operate mini-activated sludge, aerated lagoon, or DEWATS can be used to meet less stringent discharge limits. If there is sufficient space, as is often the case in rural areas, they are often followed by a constructed wetland to increase pollution removal. To meet stricter standards, a trickling filter and RBC (or each followed by a constructed wetland if there is space) are recommended.

The oxidation pond/WSP is the cheapest and easiest but is also the least effective. It includes various aerobic, anaerobic, and facultative stages, or in series. It mostly relies on solar radiation and algae to provide oxygen with sludge return to increase biological decomposition, a major difference from a typical passive lagoon treatment system. However, this can still emit NH_3 and odors and attract vectors, causing nuisances. Moreover, its anaerobic, facultative processes, and sludge drying generate methane (CH_4), a potent greenhouse gas (GHG), and nitrous oxide (N_2O) to a much lesser degree (10% of GHG emission, Vanotti et al. 2008) mainly during land application of oxidation pond sludge.

To reduce GHG emissions one option is to upgrade anaerobic digestion with an aerobic process, provided that power consumption for aeration and associated carbon dioxide (CO_2) emission off-site as well as higher CAPEX and OPEX are acceptable. With lagoons and oxidation pond/WSP, better O&M can substantially reduce GHG emission, i.e., adequate HDT, more frequent sediment dredging and sludge hauling from the ponds/lagoons with shorter intervals instead of only once in a few years (Bahia et al. 2019). Appropriate and enhanced design can also help, e.g., by adding aeration powered by solar panels, effectively turning it into an aerated lagoon.

Table 2.3 compares basic implementation requirements, associated CAPEX, OPEX, and land needed for major secondary treatment technologies, drawn from the authors' decades of experience. Depending on the material availability and equipment manufacturing capacity in each country, relative costs for both CAPEX and OPEX might differ from those in Table 2.3.

Table 2.3: Comparison of Major Secondary Treatment Technologies

Treatment process	Level of treatment	Construction difficulties		Land requirement	Cost estimates	
		Civil works	E&M		CAPEX	OPEX
Activated sludge	1	1	1	8	1	1
Trickling filter	3	3	3	6	3	3
Rotating bio-reactor	2	4	2	7	2	2
Constructed wetland	6	2	6	2	7	6
Aerated lagoon	7	7	4	3	4	5
Oxidation pond	8	8	8	1	8	8
DEWATS	4	5	7	4	5	7
Mini-activated sludge	5	6	5	5	6	4

CAPEX = capital expenditure, DEWATS = decentralized wastewater treatment system, E&M = equipment and material, OPEX = operating expense.

Note: The numbers indicate relatively high (1) to low (8).

Source: Author's experience.

As discussed earlier, wastewater characteristics dictate not only the choice of treatment technology and its designed capacity but also the CAPEX, OPEX, and the O&M complexity. A suitable primary treatment can reduce the pollution loading as much as possible at low cost, allowing a more cost-effective secondary treatment. An example (Box 2.1) well illustrates a typical situation facing many small-scale operations and the need for combining various stage and technologies, from pre-treatment to primary treatment and secondary treatment.

Box 2.1: Example of Animal Husbandry Wastewater Treatment

A livestock breeding center is estimated to generate 15 m^3/day of wastewater flow with pollutant concentrations of COD 3,000–4,000 ppm and BOD 1,500–2,000 ppm. With such high COD and BOD, it will most likely require significant pre-treatment to lower these pollutant concentrations as much as possible, such as removing solids by methods given in section 2.2. Then primary treatment described in section 2.3 is needed to remove more TSS and some BOD and COD. Typically, SMEs generate wastewater of low volume but high pollutant concentrations. Thus, a flow equalization unit is also urgently needed to stabilize influent loading and minimize shocks to downstream treatment. To meet the discharge standard of COD < 100 ppm and BOD < 30 ppm, secondary treatment is still needed, e.g., a module of TF, RBC, mini-AS, or AL and OP/WSP (if land is not an issue). Usually it is a combination of some of these in order to reduce CAPEX and OPEX.

Source: Author's experience.

Removal Rate of Treatment Technology

Pollutant removal rates are normally used as guidance for the choice of different treatment technologies and planning/initial design of pollution reduction facilities during the FS stage. However, the actual performance of wastewater treatment technologies depends on many factors (the major ones are discussed below), especially the national environmental standards, discharge limits, or other applicable standards.

Generally, removal rates are ratios of the pollutant concentrations of wastewater effluent over influent. Normally, a WWTP can handle fluctuations in flow much easier than fluctuations in pollutant concentration. In fact, wastewater concentrations fluctuate significantly for major parameters such as COD, BOD, TSS, NH_3-nitrogen, total nitrogen, and total phosphorus. As a result, their removal rates do not remain the same, and the values in Table 2.2 are just references.

A WWTP receiving higher flow and lower pollutant concentrations may have the same loading as another scenario of less flow with higher pollutant concentrations. Basically, WWTPs are designed to handle the flow fluctuations through the HDT and the biological loading through sludge retention time (SRT). Yet fluctuating wastewater loading results in different removal rates, while higher rates do not guarantee better effluent concentrations for the same or similar treatment processes. Thus, WWTP effluent concentrations commonly vary due to constant variations in loading.

For instance, when influent wastewater has a higher COD concentration than normal, well-trained operators typically can adjust rates of RAS, WAS, and blowers for more air to accommodate this. Adjusting the rates of RAS and WAS primarily provides the biological treatment system with a different functional SRT to deal with the load variation typical for WWTPs. These operational adjustments are normal and provide the biological treatment system with the capability to treat the higher pollutant concentrations, particularly for meeting required discharge limits.

In any wastewater treatment design, either centralized or on-site, biological reaction can be enhanced by providing larger volume of aeration tanks or extending the SRT to achieve better performance and so meet more stringent discharge limits. All these need to be well balanced not to create side effects such as aged sludge for bulking, which eventually impact the overall biological treatment process. Land and space availability and related investment costs are other key constraints.

The primary treatment process is typically a combination of units from influent wastewater pumping, bar screen, grit removal chamber, to primary sedimentation tank. Each unit has its own feature to remove untreatable materials prior to the secondary treatment process. In fact, their treatment efficiencies do not differ greatly from one another, mainly because the removed materials mostly are plastic, timber materials, sand and grit, and settable solids. Thus, overall pollution removal rates are primarily achieved by the secondary treatment process, which determines the ultimate performance.

In summary, removal rates can be indicative of performance and treatment efficiency of a treatment technology, and not fixed numbers for any treatment process. Each treatment system is unique, with its wastewater characteristics and applicable discharge limits. High removal rates claimed for any treatment technologies do not necessarily guarantee compliance with the applicable effluent standard. Nevertheless, they are useful parameters for judgment, especially at the FS stage of project development.

Cost Dimensions

For similar wastewater treatment processes, CAPEX can vary significantly, mainly due to equipment and material needs and associated expenses on either chemical or power supply primarily for OPEX. Generally, activated sludge processes have the highest CAPEX and OPEX, the lowest land requirements, and the best effluent qualities, and are the most reliable among all processes. In contrast, an aerated lagoon and oxidation pond/WSP require significantly more land; yet their CAPEX and OPEX are comparatively low (Sekandari et al. 2019 and Pryce et al. 2022).

The trickling filter, RBC, constructed wetland, and mini-activated sludge treatment processes have similar land requirements with good effluent quality and relatively modest CAPEX and OPEX compared to activated sludge. Regardless of the treatment process used, discharge limits can be mostly met based on effective treatment design and operational adjustment flexibility. The bottom line is a well-planned approach and innovative design in accordance with the wastewater characteristic and discharge limits. In addition, responsive and effective O&M dictate overall performance.

According to the cost estimates referenced from other similar and applicable projects as mentioned above, cost variations need to be considered, such as scale of project, availability of equipment and materials, capacity of technical support for construction, and related O&M staff and abilities. As every project is unique in the characteristics and available technologies to be chosen, the cost estimates given in Tables 2.1 and 2.2 are relative terms for comparison purposes.

Even among similar treatment processes, OPEX varies greatly, mainly due to equipment requirements and particularly the day-to-day demand on the workforce, chemicals, and electricity. In general, annual OPEX could be approximately 5%–10% of CAPEX for SMEs, based on the authors' experience. In reality, wastewater characteristics, applicable discharge limits, and land value will differ and are key determinants of actual CAPEX and OPEX.

Energy cost is probably the most influential factor for wastewater treatment, mainly because of the need for aeration and pumping, indispensable for secondary treatment, to treat wastewater of higher concentration and/or to meet stricter discharge limits. In addition, they entail sophisticated equipment and control systems, leading to more power consumption. Typically, energy requirements represent 30%–50% of OPEX, mainly comprising influent and effluent pumping, biological process aeration, associated sludge pumping, and odor control. Therefore, treatment processes need to be optimized as much as possible.

3 Discharge Standards and Implications

There are two main types of environmental standards: quality standards for environmental media such as air, water, and soil (also termed ambient standard); and discharge or emission standards that regulate the pollution released to the environment.

Ambient Versus Discharge Standards

The ambient standards define the level of environmental state in order to guarantee a basic environmental quality for all people and living resources. Thus, concentrations of their parameters are derived primarily by scientific, medical, and epidemiological research (e.g., for air quality), and eco-toxicology and susceptibility tests for aquatic life, e.g., dose–response for water quality. As a result, such standards are universal for human health and biomes. In recent years, more convergence of environmental quality standards in many countries toward international ones has occurred, such as the World Health Organization targets for air and water quality, as citizens in many countries demand a better living environment.

Discharge standards for air emission and wastewater effluent, on the other hand, regulate the maximum level of certain pollutants that a polluter can emit and/or discharge. They constitute a set of criteria for compliance by polluters, against which the regulators can check and manage pollution. In developed countries, they are designed primarily based on discharge levels economically achievable by best available technology (BAT) and experience from other countries with similar socioeconomic conditions. Technology does not just include equipment and treatment processes but also techniques and operational practices.

For example, in their national effluent guidelines, the US Environmental Protection Agency (EPA) states that "these are technology-based and intended to represent the greatest pollutants reduction that are economically achievable for an industry." It classifies pollution control technologies in the following ascending order: best conventional technology, best practicable currently available technology (BPT), and BAT. Best conventional technology and BPT are more self-explanatory than BAT, for which the definition is still vague but commonly understood as follows:

- Best—most effective and efficient, meaning when marginal cost equals the marginal benefit, resulting in overall least cost to society as a whole;
- Available—generally accessible, but not necessarily in general use yet.

Given the definition of BAT, emission and discharge standards based on it will require polluters to comply with some efforts, i.e., technical capacity and financial resources that are often lacking in developing countries. Even the US EPA only issued pollutant limits achievable by BPT (not by BAT) as national guidelines, based on which states develop and enforce their own effluent standards and discharge permits. The unique political system of the US aside, this shows the realistic approach adopted by the developed countries in setting their emission and discharge standards. The same approach has been used in Japan in setting their environmental standards. In comparison, the discharge standards used in the People's Republic of China (PRC) in the 1990s were largely copied from the former Soviet Union, and with it, the tradition of "stringent standards but lax enforcement."

As a result, emission and discharge standards can vary across states more than ambient standards due to different circumstances, and thus BPT or BAT needed to reach a similar environmental quality. Based on unofficial English translations obtained by the authors, domestic or general wastewater discharge standards of Southeast Asian countries, all in the tropics with similar socioeconomic development level, are summarized and compared with those of the US, the United Kingdom (UK), the PRC, and multilateral development banks (MDBs) in Table 3.1.

Table 3.1: Comparison of Domestic Wastewater Discharge Standards
(all max. limits in mg/L except total coliforms: MPN/100 mL)

	INO[a]	PHI[b]	VIE[c]	CAM[d]	Lao PDR[e]	PRC[f]	EHS[g]	US EPA[h]	UK[i]
BOD_5	30 (150)	50	50	80	30	60	30	53	50
COD	100 (300)	100	175	100	120	150	125	NA	250
TSS	30 (400)	100	100	80	50	150	50	30–60	NA
NH_3-N	10 (10)	NA	NA	7	NA	25	NA	NA	NA
Nitrate	NA (30)	14	TN: 40	20	NA	NA	TP: 10	NA	TN: 10–15
Phosphate	NA (NA)	1	TP: 6	6	NA	1	TP: 2	NA	TP: 1–2
O&G	NA (20)	5	20	15	5–15	15	10	38	NA
Total coliforms	3,000 (10,000)	10,000	5,000	NA	NA	NA	400	NA	NA

BOD = biochemical oxygen demand; CAM = Cambodia; COD = chemical oxygen demand; EHS = environmental, health, and safety; INO = Indonesia; Lao PDR = Lao People's Democratic Republic; NH3 = ammonia; O&G = oil and grease; PHI = Philippines; PRC = People's Republic of China; TN = total nitrogen; TP = total phosphorus; TSS = total suspended solids; UK = United Kingdom; US EPA = United States Environmental Protection Agency; VIE = Viet Nam.

[a]   In addition to limits on domestic wastewater, Indonesia also has discharge standards for general wastewater (in brackets) which refer to those not covered by its sectoral discharge limits for 48 sub/sectors.

[b]   The Philippines has four classes corresponding to surface water quality zoning. Only the class C for wastewater discharged to waters designated for agriculture use is listed to be comparable with other countries.

[c]   Viet Nam has two classes of domestic wastewater: Class A is discharged to water bodies that can be used as sources for drinking water; and Class B is for the rest, thus listed here for comparison.

[d]   Cambodia has two classes: discharge to protected water and other waters/sewer. The latter is presented.

[e]   The Lao PDR standard has two sets of limits: discharge to natural water and public water (2–3 times lax). The former is mostly used in ADB projects and thus listed here, also considering that the Lao PDR does not have WWTP yet.

[f]   The PRC's wastewater standard has three classes corresponding to five grades of surface water quality. Class 2 is for discharges to grade IV–V waters (for agriculture and industrial use) thus listed. Its centralized or municipal wastewater treatment plant (WWTP) has a different effluent standard also with several grades.

[g]   Source of EHS is Environmental, Health, and Safety General Guidelines of the World Bank. These limits are not applicable to centralized, municipal WWTPs which are in the EHS Guidelines for Water and Sanitation.

[h]   These by the US EPA are monthly averages and based on BPT, not applicable to centralized WWTPs.

[i]   This is for domestic wastewater with nitrification, which also requires reduction in total load of influent (70%–90% for BOD5, 75% for COD). Meeting either concentration or load limits is counted as compliance.

Sources: Unofficial English translation from country offices of ADB; UK and US EPA see References.

Some countries like the US only set limits on BOD, especially in early years, because the BOD/COD ratio for typical domestic wastewater is relatively stable. The total nitrogen is defined as the sum of Kjeldahl nitrogen (organic nitrogen and NH_3-nitrogen), nitrate (NO_3)-nitrogen, and nitrite (NO_2)-nitrogen. The total phosphorus is the sum of inorganic (mainly phosphate) and organic phosphorus compounds. In typical domestic wastewater, phosphate is about two-thirds of total phosphorus. That explains why some countries only have a limit on phosphate, which is easier to test than total phosphorus.

Implications of Different Discharge Standards

Table 3.1 shows that the difference in effluent standards for domestic wastewater alone ranges within 50%–100% albeit some similarities. Such a difference can greatly affect the affordability of pollution control BPT or BAT in different countries and thus cost of compliance. If polluters cannot meet the discharge standards with reasonable effort using BPT/BAT in an economically achievable way (as termed by US EPA), they just shut down the treatment system while the regulators are away or dilute effluent intentionally or unintentionally, or give up, as has routinely happened in many countries. The result is deteriorating water quality despite the stricter discharge standards or more resources to regulators. Even much more developed countries such as the US cannot afford to oversee every major polluter even using online monitoring devices.

Most countries differentiate their effluent standards by types or sources of wastewater. Some also differentiate by whether to discharge directly or into sewers connected with centralized or municipal WWTPs, e.g., Cambodia, although it is not explicit on sewers being connected to a WWTP. Only a few developing countries go further to differentiate their discharge standards by quality of receiving waters, like the PRC and the Philippines. Waters in classes less demanding on water quality, e.g., those for industrial and agriculture use, can tolerate less stringent effluent limits in general. Waters in the most stringent (cleanest) class do not allow any wastewater discharge, as is the case for the PRC.

This cascading approach has several benefits. First, it can overcome the lack of technical capacity and reduce the burden of individual polluters, especially smaller ones common in developing countries. They just need to meet the standards for discharging wastewater into sewers that are connected to a WWTP. Such a standard thus has less stringent discharge limits than those for direct discharge to the adjacent waters or soil. Combined with the economy of scale of a WWTP, it is more cost-effective for the system and society as a whole. However, many developing countries, especially smaller ones, do not have or have just started to build municipal WWTPs. Hence, their standards are less finely divided into classes that have separate, less strict limits for discharging to sewer–WWTP.

Second, this approach can reduce the incentive for diluting the discharge and thus save water and foster cleaner production practices. Undiluted wastewater is more desirable for WWTPs, which function better at higher concentrations with less energy used per pollutant reduction. Typical sewage WWTP influent is around BOD 200–300 mg/L especially in temperate zones. This becomes a major standard (BOD < 300 mg/L) for discharging wastewater into the sewer networks of a WWTP in the PRC, and BOD for direct discharge to class IV–V waters (for industry and agriculture use) is less than 60 mg/L.

Therefore, it is recommended that developing countries set their ambient and discharge standards in categories or classes according to main uses and correspond them to each other. Effluent standards should be divided according to the quality of receiving waters suited for their designated uses, tapping the self-purification capacity of surface water (e.g., assimilation, degradation, and hydrolysis). The numeric limits in these standards should be established mainly based initially on what is economically achievable by BPT, and by BAT as a country improves in capacity and financial resources. All these require research and experience often beyond the reach of many DMCs. Therefore, they could introduce such standards by borrowing from countries with similar circumstances first and gradually adapt them to their own situation.

From Concentration-Based Standard to Total Load Control

With population and economic growth, even if most polluters along a river or lake or airshed comply with the discharge standards (predominantly concentration-based), the water quality of that river/lake or air quality in that area can still worsen. The first response is tightening the discharge standards—often to no avail after a few years. This is because total pollution load discharged to the water body or airshed in question is beyond its assimilative capacity to absorb and degrade. Therefore, total pollution load control is also needed.

Environmental authorities in developed countries like the UK and the US require compliance both in concentration and total pollution load for a certain period when issuing discharge permits, in order not to exceed the carrying capacity of the receiving environment. Similarly, the EIA in Indonesia and the PRC also requires estimating pollution load in addition to discharge concentration in order to demonstrate compliance (or not) in both. This raises the question of how to determine carrying capacity of a water body or airshed (thus the maximum amount of certain pollution allowed during a certain period).

In the PRC, this has been attempted through environmental planning since the mid-1990s for heavily polluted rivers. Learning from developed countries, the first step is to estimate the maximum total load (e.g., in terms of COD amount) that can be absorbed by the river or its segments while maintaining its water quality at a certain level. Geographic, climatic, aquatic life, topographic, and hydrodynamic factors can affect a river's capacity to decompose and degrade pollutants. With so many factors in play, as well as the complexity and scientific uncertainty involved, it is easy to see the resulted total pollution load or carrying capacity estimate can be controversial and disputable.

In the next step, the planners allocate the estimated total pollution load considered absorbable by the water body to administrative units in its catchment, mainly on a grandfathering basis. Each locality, in turn, allocates its share of total pollution load among its key polluters across sectors. Localities and sectors that have already exceeded their share of total load must shut down some heavy polluters beyond repair while balancing social, economic, and environmental obligations (e.g. not big employers or contributors to tax revenue). This has added push for restructuring of the economy to being less pollution-or-resource-intensive and technical upgrading among industries.

However, the above approach has mainly tackled point-source pollution, as evident in the limited improvement in water or air quality. In the PRC, studies revealed that water pollution comes more from non-point sources such as agriculture runoff than point sources like urban sewage and industries, roughly in the ratio of 2:1. The MDBs also noticed this after years of investment in municipal WWTPs. As a result, some of their projects adjusted the objectives from improving water quality of a river to reducing pollution load discharged into it. Therefore, the environmental master plans need to begin with pollution source analysis beyond point sources to cover all major sectoral contributors in the geographic unit in question.

Results of pollution source analysis and the cost–benefit analysis of their abatement, even if preliminary, can be more powerful than awareness campaigns, training, and workshops with regard to convincing the government and the public. An environmental master plan developed in this way has the following benefits:

- A more systematic approach and recipe with maximum benefits and minimum cost for a water or airshed or jurisdiction, not piecemeal as often seen in some projects and initiatives that picked seemingly good technology for a sub/sector, which may turn out not to be the most cost-effective and thus unsustainable or hard to replicate.

- Quick tangible results in environmental quality improvement by undertaking first the most cost-effective pollution control (lower hanging fruits) can boost public confidence and garner government support for more substantial investment, also gain experience to tackle tougher pollution and sub/sectors later.

- On the practical side, an environmental master plan developed through source analysis and cost–benefit analysis is well-reasoned by definition and can offer a road map on how to do what and when. Its investment proposals can thus be more easily justified to convince financiers (e.g., MDBs and various funds) and facilitate their engagement.

4 Technologies for Solid Waste Management

Principles and Waste Management Hierarchy

Typical solid waste of livestock husbandry includes (1) dung (e.g., cow pats or manure) and undigested residue of plant matter; (2) waste or uneaten feed; and (3) soiled bedding material (e.g., straw, sawdust, wood shavings, and paper-based bedding materials). Solid waste of aquaculture typically includes (1) sludge from feeding (e.g., from fish farms or shrimp ponds); (2) shrimp shells and fish scales; and (3) tissues and skins of fish including unwanted parts.

Solid waste from agro-processing is more diverse. Those from slaughtering and meat processing mainly include blood, organs, manure/litter, residues of bedding and feed, feathers, and hatchery waste. Those from plant-based processing are mainly peels, trims, cuts, fruit stones and shells, and husks. Those from fish processing mainly include heads, bones, shells, and other inedible offcuts. Their amount as percentage by weight of raw material or finished products is high, with a range of 30%–60% or higher (see the study on pollution characteristics for agriculture subsectors under the same TA).

The waste management hierarchy begins with waste avoidance and minimization, followed by reuse and recycling with the residues for safe disposal. On waste minimization at source and during the process, the International Finance Corporation/World Bank EHS guidelines on sub/sectors in this study provide major measures and operational practice. Recycling and reuse include both of material (i.e., water, nutrients, and organic content, such as by composting or digestion) and energy content (e.g., through drying, pyrolysis, and incineration). Disposal methods in agriculture include burning or incineration, landfilling, and disinfected safe burial.

Similar to liquid waste/wastewater management, solid waste segregation should begin as early as possible, preferably before mixing with liquid wastes, which in some cases like animal pens is unavoidable. This timely action not only reduces the moisture content in collected solid waste, thus improving recyclability and treatability, but also lowers the quantity. In addition, as agro-sectors are dominated by organics that can be readily utilized, they should be separated from other inorganics and hazardous wastes. Therefore, analyzing their characteristics is the top priority since the result will dictate the choice of technical options, their designed capacity, cost to build and operate, and O&M.

Hazardous wastes are defined as those that possess the following properties: toxic, infectious, pathogenic, flammable, corrosive, or reactive. The abovementioned sectoral EHS guidelines describe typical hazardous wastes from different sub/ sectors. Their characteristics and management are much more complicated than for regular wastes. Their small amounts combined with a lack of economy of scale at SMEs further jeopardize proper management. Therefore, it is imperative to minimize hazardous waste first and try to convert it into non-hazardous waste if possible and as permitted by health and safety requirements, e.g., by thermal disinfecting infectious and pathogenic wastes (Ren 2022).

Experience worldwide shows that the key for success of any recycling, in particular of organics highly relevant for agro-wastes, is to minimize impurity in advance, i.e., by robust waste sorting and pre-treatment. Impurities can interfere with the biological processes in composting and digestion, affect the quality of recycled products (e.g., compost), and thus their application. Inferior uses and application, in turn, fetch a lower price than the recycled products would otherwise. Limited application or markets for recycled products and lower prices, both due to their inferior quality, undermine long-term sustainability of recycling, composting, and digestion. With adequate pre-treatment, the main process will perform better and also reduce unnecessary loading from pre-treated waste.

Solid Waste Management Technologies

Depending on the waste characteristics and financial capacity of the project several major technologies are available for organic waste management and utilization suitable for small scale in a rural setting. The most simple and common practice is piling on-site, with digestion or fermentation naturally occurring that turn waste into fertilizer for farmers to use. However, for animal farms of the scale for typical cooperatives or even bigger as SMEs, such practice is unable to properly manage waste, especially manure, causing contamination of water, odor nuisance, and hygiene hazards. The seasonality in fertilizer use and thus demand by farmers further exacerbates the situation, all contributing to animal husbandry being the top polluter in rural areas of many countries.

Solar and wind drying

The next slightly complex method for organic waste, especially manure, is **natural drying** by sun and wind. This has historically been the conventional method in arid areas, where the quickly dried manure is collected and stored for use as fuel in winter or in warming up animal pens or other facilities.

In tropical and wet climate zones, natural drying faces more challenges but with some success, e.g., by a combination of solar and wind drying. The solar drying method functions well but heavily depends on the waste characteristics, climate conditions, land availability, and careful design of the facility. With a shed designed to enhance ventilation or mechanically assisted with low power consumption and shelter from the weather (e.g., rain) and odor control, solar drying technology offers a good conversion of waste to a resource as well as cost-effectiveness for both CAPEX and OPEX (Sheng et al. 2022).

Composting

Composting is an aerobic method of recycling organic content by decomposing organic solid wastes into a humus-like end product known as compost that can be used as soil conditioner and fertilizer. It is a relatively fast biodegradation process, normally taking about 4–6 weeks to reach waste stabilization (see Figure 4.1). The composted wastes are odorless with better texture and low moisture content. In general, factors suitable for the composting process include a temperature of 52–60°C, moisture content of 50%–60%, pH of 5.5–6.8, and oxygen content of 15%–20%.

The major composting methods include passive piles, windrow composting, static piles, and in-vessel composting (in bins, beds, silos, transportable containers, and rotating drums). Passive piles require more land and a longer time, while in-vessel composting is more costly both CAPEX and OPEX for agro-sectors.

Windrow composting involves placing mixed materials in long, narrow piles and turning or agitating them regularly. This is the most common method used for rapid composting of yard wastes. Windrows are formed using a front-end loader, and are turned with this equipment or a specialized turning device. Windrow turning frequency depends on decomposition rate, moisture content, and porosity of the materials, and the desired composting time. To speed up composting, turning windrows once or twice daily for the first week is necessary and thereafter every 3–5 days.

Typical benefits of the composting process include (1) adding organic matter, improving soil structure, reducing fertilizer requirements, and potentially reducing soil erosion; (2) decreasing the volume and moisture content but increasing stability of manure; (3) easier to handle manure without odor or fly problems, lowers pollution and nuisance; (4) high-carbon manure and bedding mixtures lower the carbon/nitrogen ratio to acceptable levels for land application; and (5) appropriate temperature within the compost piles to reduce pathogens.

Biodigestion or anaerobic digestion

Biodigestion or anaerobic digestion is a biological decomposition of organic matter by bacteria in the absence of oxygen (see Figure 4.1). It produces biogas consisting of mainly CH_4 (55%–65%) and so can be used as fuel. It can kill pathogens and thus make the bio-slurry and sludge safer for land application than conventional simple piling and fermentation. Indication of well-functioned digestion is an odorless product, as much of the original nitrogen is retained in the liquid fraction.

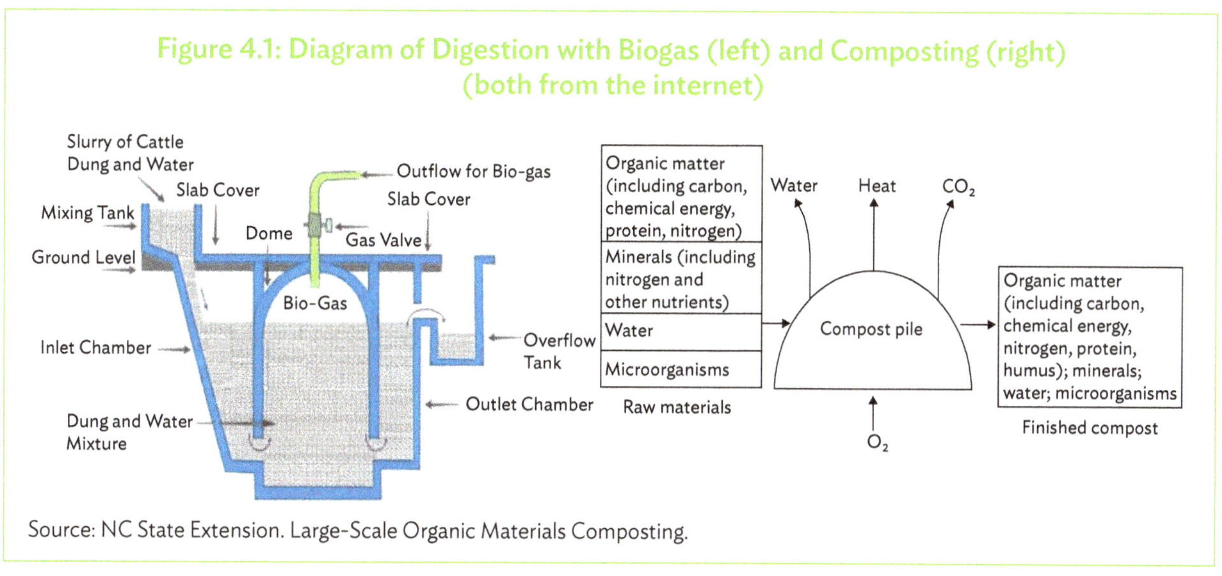

Source: NC State Extension. Large-Scale Organic Materials Composting.

Biodigestion with biogas entails piping, valves, gas purification, and sulfide removal devices, stoves, and lighting designed for using biogas, which has lower heat value than natural gas. All require additional investment, technical service for repair, and training to users albeit its obvious benefits in supplying cheap and cleaner energy supply, improving indoor air quality, and reducing GHG emissions.

As mentioned earlier, the demand for fertilizer and thus the use/sale of organic recycled products fluctuates by season whereas the agro-wastes such as manure are generated daily. In this regard, composting has some advantages over drying and digestion, as its end product, compost, is generally drier (than digestion) and cleaner (than drying), enabling longer shelf life, ease of transport over long distance, and thus better application and sale. A comparison of these technologies for organic wastes are summarized in Table 4.1.

Table 4.1: Major Organic Waste Technologies Applicable to SMEs

Items	Solar/wind dry	Composting	Biodigestion/biogas	Waste-to-energy
Pros	Simple, inexpensive, easiest in O&M	Faster, less costly than biodigestion, easier O&M	Save energy cost by using CH_4, reduce GHG emission	Reduce volume up to 90%, highly effective pathogen destruction
Cons	Low in destroying pathogens, odor, attracts flies and rats, etc., hygiene issue	Higher demand in O&M to effectively destroy pathogens, small amount wastewater, hygiene issue	Higher demand in O&M to effectively destroy pathogens, small-scale may not be worth CAPEX and OPEX for biogas	Expensive, complex O&M, toxic flue gas (dioxin, etc.) and fly ash requiring costly treatment, monitoring
Waste reduction	Medium/low	Medium	Medium	High
CAPEX	Low	Medium	High (w. biogas) /medium	High
OPEX	Low	High/medium	High/medium	High
O&M	Easy	Moderate	Moderate	Highly complex

CAPEX = capital expenditure, CH4 = methane, GHG = greenhouse gas, O&M = operation and maintenance, OPEX = operating expense, SMEs = small and medium-sized enterprises.

Note: With very little quantitative information available and for comparison purposes, pathogen removal, waste volume reduction, CAPEX, and OPEX, etc. are all in relative terms.

Sources: Sheng et al. (2022) and the authors' experience.

Technologies for Hazardous Waste Management

In many countries, hazardous wastes and substances are those that exhibit at least one of these properties: toxic, flammable, reactive, pathogenic, infectious, corrosive, and radioactive. Toxic wastes include many residues of chemicals, biochemicals, antibiotics, and other drugs widely used in agro-sectors and their laboratories. Infectious waste contains pathogens (e.g., bacteria, viruses, parasites, or fungi) in sufficient quantity to cause disease to susceptible hosts. Such waste includes dead or sick animals, test animals, tissues, microorganisms cultures, and their containers. There are also other general hazardous wastes not unique to agro-sectors but common in all industries and operations, such as batteries, waste oil, and solvents.

The last group of general hazardous wastes often needs to rely on off-site treatment by specialized entities, with prior collection and storage on-site. The study focuses on methods suitable for on-site and small-scale hazardous waste management. They mainly include disinfection by chemicals (e.g., lime), by autoclave or similar wet thermal treatment, and microwave, as well as final disposal by burial or incineration.

Disinfection or sterilization is used to convert infectious wastes into largely noninfectious general wastes, disposal of which is much easier and cheaper in complying with EHS requirements. Disinfection by chemicals is conventional yet effective, e.g., lime is used to treat dead or culled animals before their final burial. The relative efficacy in removing hazardous properties through main disinfection methods are compared in Table 4.2.

Table 4.2: Major Technologies for Hazardous Waste from Agro-Sectors

	Disinfection			Final disposal		
	Autoclave	Microwave	Chemical	Pyrolysis	Incineration	Safe burial
Pros	Efficacy high, standard tech., simple, fast, high penetration	More energy / time efficient, low pollution, easier to control, fast	Widely applied, simple, low temperature	Less flue gases incl. dioxin, has fuel gases (CO, CH_4), oil and bio-char byproduct	Widest use, reduce waste up to 90%, mature tech.	Simple, cheap and convenient, traditional
Cons	High temp and pressure, soft water, emit gases and wastewater	Similar/lower efficacy as autoclave, need pre-treatment	High cost in agents, residue wastewater and chemicals	Energy intensive thus expensive, need pre-drying, grinding, emit PAHs, HCl	Toxic flue gases (esp. dioxin, furan, mercury) and fly ash	Contaminate water and soil, hygiene nuisance (rats, flies)
CAPEX	Low	Medium	Medium	Medium-high	High	Low
OPEX	Low-medium	Low	Medium	High	High	Low

CO = carbon monoxide, CH4 = methane, HCl = hydrogen chloride, PAHs = polycyclic aromatic compounds.

Sources: Kollu et al. (2022), Pariha et al. (2019), Reyes et al. (2015), Su et al. (2021), Tony et al. (2011), IFC (2008), and the authors' experience.

For chemical hazards and toxic wastes, common methods include neutralization by basically mixing waste acids and alkali solutions or substances, e.g., those from laboratories. Stabilization by various chemical means, e.g., oxidization and reduction, and inertization by physical means like encapsulation by concrete or asphalt, or mixing with fly ash to form a cement-like solid mass that resists leaching.

As final disposal, burial or landfilling are conventional methods that, if properly done, remain easy, cheap, and effective. The siting, proper design, and construction are crucial to prevent and minimize contamination of groundwater and soil. Location and site conditions with minimal permeability (e.g., clay soil and distance to bedrock, aquifer, and other water sources) are key requirements that can lower the cost by reducing the need for additional treatment or liners to meet standards on non-permeability.

Incineration is commonly used disposal for hazardous wastes in developing countries. For infectious wastes from veterinary facilities, laboratories, the EHS guideline for health facilities recommends two-stage incineration with the first stage using pyrolysis, an air or oxygen-deficient medium-temperature combustion (800–900°C). The fuel gases produced in the pyrolytic chamber are burned at high temperature (900–1200°C) for further thermal decomposition. This is called pyrolytic incineration, also known as controlled air or double-chamber incineration.

However, it is well known that incineration of chlorinated plastics and solutions, etc. unavoidable in infectious wastes, and incineration of garbage can generate particulate matter, heavy metals, and dioxins in flue gas emissions and bottom ash. Many industrialized countries are phasing out incinerators for such waste. Although pyrolysis emits less flue gases, it has its share of issues and risks (Table 4.2). As an endothermic process, as opposed to incineration which is exothermic, pyrolysis is usually more energy intensive. It also requires pre-treatment of waste (drying, shredding, or grinding), all energy intensive and quite expensive, contributing to its much lower adoption than incineration.

5 Technologies for Air Emissions Control

Air emissions from agro-sectors mainly include odor [e.g., NH_3, hydrogen sulfide (H_2S), and thiols] from various types of production and processing, and CH_4 and N_2O (GHGs) from animal husbandry. These are mostly non-point source fugitive emissions, as compared to typical flue gas emissions from combustion (e.g., boilers) which are point-source air emissions. The study focuses on fugitive air emissions of agro-sectors, since there is a wealth of information on flue gases from combustion.

Principles

Similar to liquid and solid waste treatment principles, collection mechanisms for air emission and odor should be provided as near the source as possible, so as to minimize the size of downstream treatment. This effective collection will not only lower the CAPEX and OPEX but also significantly minimize the environmental impact. In addition, temperature and moisture of the collected are generally major factors influencing the treatability and efficiency of the available treatment technologies. Thus, good management of such pre-treatment is desirable for more efficient air emission control.

Air Pollution Control Technologies

Depending on the emission limits and characteristics of air emission and odor, a combination of two or more treatment processes may be required. The most commonly used physical method is activated carbon adsorption or liquid washing, while the chemical method is chemical washing, flushing, and/or combustion or incineration.

For SMEs, neither combustion nor incineration are viable options considering their high CAPEX and OPEX. Biological and chemical methods include flushing/washing, stripping, and the application of biofilm. Considering the characteristics of agro-sectors and their air pollution, three types of control technologies are recommended for their cost-effectiveness and are briefly discussed below.

Activated Carbon

Activated carbon is an adsorbent commonly used in the treatment of low concentrations of air emission and odor. This technology typically involves adsorbing organic matter onto a porous solid surface of carbonized material to remove odor. Activated carbon is usually obtained from waste products such as coconut husks and other agricultural wastes, then converted into charcoal before being "activated." It is a form of carbon commonly used to filter pollutants from water and air, among many other uses. It is processed (activated) to have small, low-volume pores that increase the surface area available for adsorption effect.

Filters with activated carbon are usually used in compressed air and gas purification processes to remove oil vapor, odor, and other hydrocarbons from the air. The most common design is to apply a one- or two-stage filtration where granular activated carbon (GAC) is embedded inside the filter media. The filter beds with GAC adsorb air and odor compounds while the filtered air passes through to the exhaust tower or chimney.

Temperature (below 40°C) and moisture are key factors for a well-managed GAC system. In addition, dust, smoke, and impurities can also affect adsorption, thus necessitating certain pre-treatment. Also, minimizing the collection area, such as using a hoop is necessary. The used GAC can be regenerated by heating at high temperature in a furnace or oven. Such treatment and final disposal can be costly because used GAC can become hazardous waste.

Air Stripping

In this technology, air emissions are put into contact in a stripping tower packed with media with chemical liquid sprayed down to transfer volatile organic compounds and odorous substance from the gas phase to the liquid which can then be treated through neutralization, oxidation, and/or other reactions. Air emissions that can be stripped include organic sulfides, nitrogen-containing compounds, organic acids, and a few hydrocarbons. Generally, alkaline and acidic odor components can be neutralized using related chemical solutions, respectively. This method can only convert odorous molecules into salts for adsorption.

Normally, air stripping equipment operation is affected by algae, fungi, bacteria, and fine particles, especially on the media within the stripping tower, so it requires periodic cleaning. Air stripping is effective only for air emissions with volatile organic compounds or semi-volatile compounds with a Henry's Law constant greater than 0.01. Compounds with low volatility at ambient temperature may require preheating and thus entail extra cost. Overall, air stripping can be challenging for SMEs due to its relative high CAPEX and OPEX and complex O&M.

Biofilter

Biological treatment of air emissions has the advantages of being economic, limited secondary contamination, and high efficiency. The bio-filter and bio-trickling methods have been most used recently. Owing to relative simplicity, long usage, and user-friendly mechanism, bio-filters are highly recommended for SMEs. They transfer organic matter from the gas to a liquid or solid phase biofilm, which is absorbed by microorganisms and then oxidized into CO_2 and water. In addition, biological enzymes have the catalytic decomposition ability to break down odor molecules and inhibit the growth and reproduction of spoiling bacteria.

However, due to the speed of gas loading and the limitations of the bio-filter system, its effectiveness in pollution removal is a bit lower than chemical methods (e.g., air stripping). Like other biological processes for wastewater and solid waste treatment, bio-filters for air emissions are inherently harder to handle than physical–chemical processes like activated carbon and chemical stripping. The treatment agents, microbes, are more susceptible to pH, temperature, flow, type of pollutants, and their concentration.

Similar to wastewater and solid waste treatment, quantity and quality of air emissions and emission standards applicable to SMEs, investment of CAPEX and OPEX, as well as the complexity of O&M are major determinants for the selection of abatement technology. Basic parameters for comparison of applicable technologies for air emission control are summarized and presented in Table 5.1.

Table 5.1: Air Emission Control Technologies Applicable to SMEs

	Activated carbon	Air stripping	Bio-filter
H_2S removal (%)	85–90	>97	91.9–96.0
NH_3 removal (%)	85–90	>97	91.0–99.5
CAPEX	Medium/low	High	Medium/low
OPEX	High	Medium/low	Low
O&M complexity	High/medium	High	Medium/low

CAPEX = capital expenditure, H_2S = hydrogen sulfide, NH3 = ammonia, O&M = operation and maintenance, OPEX = operating expense, SMEs = small and medium-sized enterprises.

Note: Due to very little information available, H_2S and NH_3 removal, CAPEX, OPEX, and O&M complexity are in relative terms for comparison purposes.

Sources: Fang et al. (2004), Tay et al. (2004).

6 Conclusions and Recommendations

A repeated theme throughout is the pollution management principles and hierarchy. The most important is separating liquid or wastewater from solids in order to control and utilize cost-effectively. Within liquid or solid states of pollution, segregation of streams of different nature and strength is needed to enable effective reuse, recycling, and abatement.

Effective segregation often entails additional investment and change in operation practice, as showcased in different cleaning methods of manure in animal holding facilities. Economy of scale can foster the adoption of more advanced methods. Many factors are at play for smallholders who dominate in developing countries, which is beyond the scope of this study. However, many cheaper options that mostly require simple changes in operation practice are readily available in many guidelines on EHS or cleaner production.

For water use, minimization and reuse should be promoted as long as permitted by hygiene and food safety requirements. This not only saves valuable water resources but also leads to less wastewater and less dilution. Both these, in turn, can greatly reduce the burden and cost of wastewater treatment, leading to better affordability and sustainability.

The discussion on wastewater illustrates the critical role of environmental standards, in particular discharge standards. The choice of pollution control technologies, central to this study, is essentially based on their likelihood of cutting down the pollution to meet applicable discharge standards. As countries develop, a more fine-tuned design based on BPT or BAT can facilitate rather than inhibit compliance by polluters, especially those small and medium sized. Yet, compliance with concentration-based standards is just the first step and foundational for safeguarding the environment.

The most salient indicator for technologies' ability to abate pollution is their removal rate of targeted pollutants, i.e., key parameters regulated by discharge standards. However, the discussion in Chapter 2 shows that other factors, notably retention time and microbe strength, also greatly affect the final results of wastewater treatment. Therefore, the removal rates of major technologies serve more as starting points and reference. With them, project designers/FS and EIA preparers can at least make judgments about validity and feasibility of various technologies proposed to them in controlling the pollution in question.

The well-known solid waste management hierarchy indicates the order of preferred actions environmentally and financially. For agro-wastes that cannot be avoided or minimized, the study focuses on technologies to recycle their nutrient content. As part of a circular economy or regenerative agriculture, using bio-fertilizer from agro-wastes can bring many benefits in addition to waste management and direct pollution control:

- Improve soil health, promote nitrogen fixing and carbon sequestration through better microbe and invertebrate populations in soil.
- Increase yield and resilience to drought with more developed root systems, contributing to food security that ADB is committed to support.
- Reduce the demand and cost for chemical fertilizers, leading to GHG mitigation.
- Enhance plants' resilience to pests and diseases, which in turn can reduce pesticide use, benefiting biodiversity and climate change mitigation and adaptation.
- Reduced use of pesticides and chemicals will generate less runoff, thus benefiting water quality and ocean health.

However, one of the biggest challenges for these technologies to utilize and recycle organic wastes is the acceptance of their end products, e.g., as fertilizer, compost, or bio-slurry from digestion. Only if they can be widely used, can they produce good revenue to sustain these organic recycling operations. This greatly depends on the quality of recycled products, which in turn demands better waste segregation to minimize impurities, the top factor making or breaking organic recycling.

As shown in the discussion in most chapters, different technologies work best for certain types of pollution in a certain range of concentrations. Their scale or designed capacity also needs to match the measured or estimated pollution amount with some margin. A lack of knowledge of pollution characteristics (e.g., volume/amount, major pollutants and their concentration range, and nature of wastes) will result in an unnecessary or useless treatment system. Therefore, results of a parallel study on pollution characteristics for agro-sectors should be used in tandem with this report.

Major Technologies for Small-Scale Wastewater Treatment

Based on literature (e.g., Bachi et al. 2022, Waqas et al. 2023, Zhang et al. 2015, Hassard et al. 2015), treatment technologies suitable for wastewater typical of agro-subsectors are summarized with their processes, and their pros and cons briefly discussed. The diagrams used are mostly from the internet except otherwise indicated.

Activated Sludge

Developed in the late 19th century, activated sludge is a process for treating domestic and industrial wastewater, which became popular in the early 20th century. It is the most commonly used wastewater treatment process for various purposes as it can achieve one or several of the following:

- oxidizing biodegradable matter, mainly carbonaceous and nitrogenous (mostly ammonium and nitrogen in biological forms);
- removing phosphate ;
- generating biological flocs that are easy to settle; and
- generating liquid low in dissolved or suspended solids.

In general, the activated sludge process consists of three treatment units: (1) an aeration tank serves as biological reactor; (2) a secondary sedimentation tank separates the settled solids and treated wastewater; and (3) pumping equipment transmits settled solids from the secondary sedimentation tanks to the inlet of the aeration tank as the returned activated sludge (RAS). The mixture of wastewater and biological mass is commonly known as mixed liquid suspended solids (MLSS) with typical concentrations of 3,000–4,000 mg/L.

Due to natural biological growth, solids eventually accumulate beyond the desirable level of MLSS concentration in the aeration tank. This excessive amount of solids, WAS, needs to be removed from the process to keep the food to biomass ratio within the designated range. The WAS is typically stored in sludge holding tanks with adequate mixers for further treatment, such as mechanical dewatering or aerobic/anaerobic digestion to reduce volume prior to final disposal.

What actually happens is that MLSS concentration increases in the aeration tank and then further concentrates through the secondary sedimentation. Also, increasing hydraulic retention time in the aeration tanks provides higher MLSS which will increase the sludge retention time so to reduce concentrations of biological oxygen demand (BOD), chemical oxygen demand, total nitrogen, ammonia (NH_3) nitrogen, and total phosphorus.

The removal efficiency of the activated sludge process is controlled by different factors, e.g., sludge retention time (i.e., aeration tank volume divided by the influent flow rate), MLSS concentration, and RAS/WAS rates. Other factors include influent wastewater loading [e.g., total suspended solids (TSS), BOD, chemical oxygen demand, nitrogen, and phosphorus] in relation to the MLSS present in the aeration tank (food to biomass ratio), air (oxygen) supply, and temperature.

The characteristics of influent wastewater vary due to wastewater generation on a daily basis. For example, typically there are two peak inflows during a 24-hour period for a municipal wastewater treatment plant (WWTP) operation. The first peak normally incurs around midday, and the second peak is at nighttime around 8:00 p.m.–10:00 p.m., mainly owing to dinner preparation, dishwashing, showering, and other night activities. Inflows become much lower after midnight until early morning of the next day. For a small treatment plant, an equalization tank can function as a savior of the entire operation by stabilizing and adequately mixing the influent.

The mixing pattern is critical since it affects the biomass loading and oxygen transfer required in the aeration tanks as well as the kinetics governing the treatment process. Mixing patterns for the activated sludge process are mostly plug flow and complete mixing. Mixing, pumping, and aerating all consume energy, which is the biggest cost item for a WWTP.

Trickling Filter

A trickling filter is a type of wastewater treatment system consisting of a fixed bed of rocks, coke, gravel, or plastic media. Sewage or wastewater flows downward to these media and provides conditions for a layer of microbial slime (biofilm) to grow and surround the bed of media. Typically, aerobic conditions are maintained by splashing and diffusion either by forced-air flowing through the bed or natural convection of air given the filter medium is porous.

Major elements of a complete trickling filter system are

- a bed of filter medium on which a layer of microbial slime is developed,
- an enclosure or a container that houses the bed of filter media, and
- a system for distributing the flow of wastewater over the filter media.

Typically, settled sewage flow enters at a high level and flows through the primary settlement tank. The supernatant from the tank flows into a dosing device, often a tipping bucket that delivers flow to the arms of the distributor. The flush of water flows through the arms and exits through a series of holes pointing at a downward angle. This propels the arms around, distributing the liquid evenly over the surface of the filter media. Unlike the diagram in Figure A1, most trickling filters are uncovered and freely ventilated to the atmosphere.

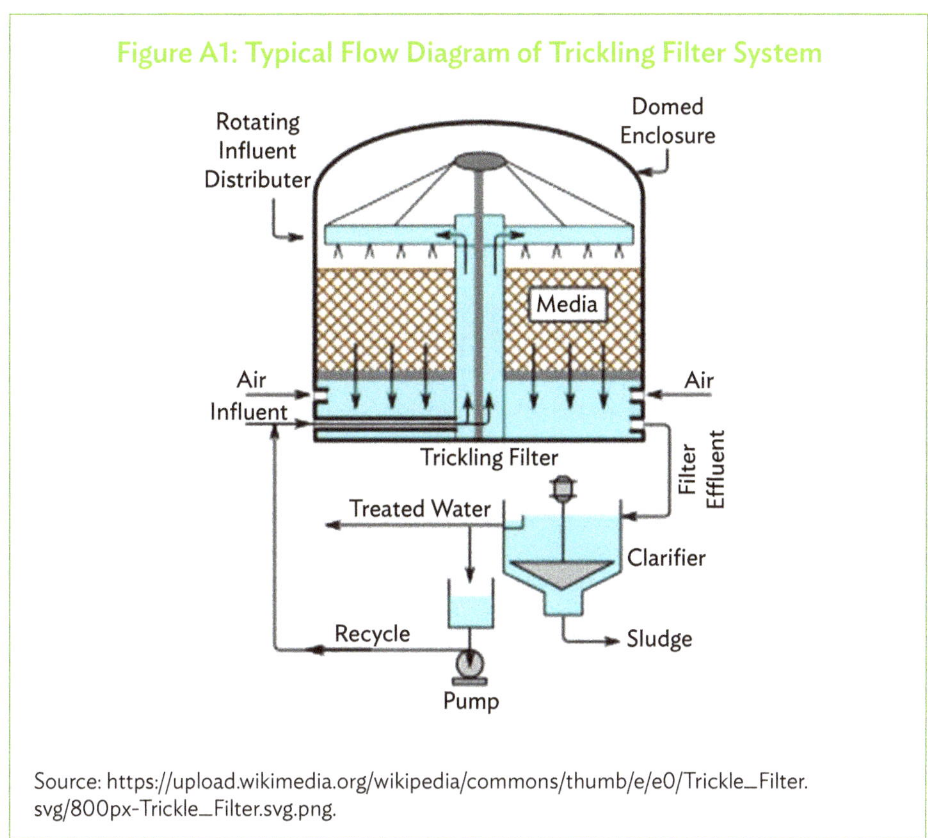

Source: https://upload.wikimedia.org/wikipedia/commons/thumb/e/e0/Trickle_Filter. svg/800px-Trickle_Filter.svg.png.

The removal of pollutants from the wastewater stream involves both adsorption of organic compounds and some inorganic ones (such as nitrite and nitrate ions) by the layer of microbial biofilm. The filter media is typically chosen to provide a very high surface-to-volume ratio. Typical materials are often porous and have considerable internal surface area, in addition to the external surface of the medium. Passage of the wastewater over the media provides the dissolved oxygen required for biochemical oxidation.

The biofilm that develops in a trickling filter may become several millimeters (mm) thick and is generally a gelatinous matrix that may contain many species of bacteria. This is very different from many other biofilms, which may be less than 1 mm thick. Normally, both aerobic and anaerobic zones can exist in the biofilm supporting both oxidative and reductive biological processes. At certain times of year, especially spring, rapid growth of organisms may cause the film to be too thick. Then, it may be cast off to become sludge to be removed, e.g., by air sedimentation tanks for further removal of pollutants and thickening of the sludge.

A typical trickling filter is circular in shape with diameter of 10–20 meters (m) and depth of 2–3 m. A circular wall, often of brick, contains a bed of filter media that, in turn, rests on a base of under-drains. These under-drains function both to remove liquid passing through the filter media but also to allow the free passage of air up through the filter media. Mounted in the center over the top of the filter media is a spindle supporting two or more horizontal perforated pipes that extend to the edge of the media. The perforations on the pipes are designed to allow an even flow of liquid over the whole area of the media and are also angled so that, when liquid flows from the pipes, the whole assembly rotates around the central spindle. Settled sewage is delivered to a hopper at the center of the spindle via some form of dosing mechanism, often a tipping bucket device on small filters.

Systems can be configured for single-pass use where the treated water is applied to the trickling filter once before being disposed of, or for multi-pass use in which a portion of the treated water is cycled back and re-treated via a closed loop. Multi-pass systems result in higher treatment quality and assist in lowering total nitrogen levels by promoting nitrification in the aerobic media bed and de-nitrification in the anaerobic septic tank. Some systems use the filters in two banks operating in series so that the wastewater has two passes through a filter with a sedimentation stage between the two passes. Every few days the filters are switched round to balance the load. This method of treatment can improve nitrification and denitrification since much of the carbonaceous oxidative material is removed on the first pass through the filters.

After several decades of operation, trickling filter now use a variety of types of filter media to support the biofilm. Types of media most commonly used include coke, pumice, plastic matrix material, open-cell polyurethane foam, clinker, gravel, sand, and geotextiles. Ideal filter medium optimizes the surface area for microbial attachment and wastewater retention time, allows air flow, resists plugging, is mechanically robust in all weathers, and does not degrade. Recent trickling filter technology involves aerated bio-filters using plastic media in vessels plus blowers to provide air at the bottom of the vessels.

Rotating Biological Contactor

While TF's biofilm is fixed, a rotating biological contractor (RBC) consists of a series of closely spaced, parallel disks covered with biofilm and mounted on a rotating shaft supported just above the surface of the wastewater. Microorganisms grow on the surface of the rolling disks, where pollutants are biodegraded gradually.

An RBC is capable of withstanding surges in organic load. To be successful, microorganisms need both oxygen to live and food to grow. Oxygen is obtained from the atmosphere as the disks rotate. The microorganisms grow and build up on the media until they are cast off due to shear forces provided by the rotating disks. The rotating packs of disks (known as the media) are contained in a tank or trough and rotate at 2–5 revolutions per minute. Commonly used plastics for the media are polyethylene, PVC, and expanded polystyrene. The shaft is aligned with the flow of wastewater so that disks rotate at right angles to the flow, with several packs usually combined to make up a treatment train.

The disks consist of plastic sheets 2–4 m in diameter and up to 10 mm thick. Several modules can be installed in parallel or in series to meet the treatment requirements. The disks are typically submerged in wastewater to about 40% of their diameter.

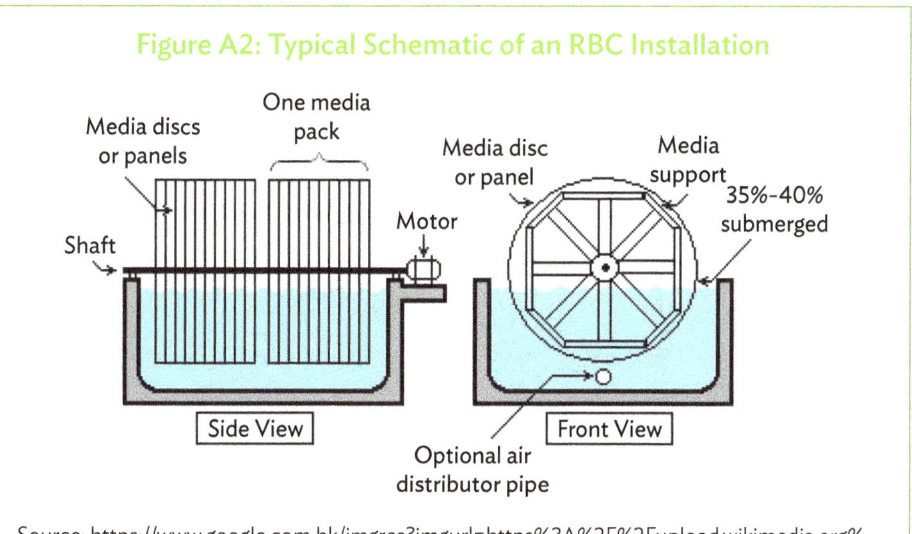

Figure A2: Typical Schematic of an RBC Installation

Source: https://www.google.com.hk/imgres?imgurl=https%3A%2F%2Fupload.wikimedia.org%2Fwikipedia%2Fcommons%2F3%2F3c%2FRotating_Biological_Contactor.png&tbnid=LOclvBb5DZ8TPM&vet=1&imgrefurl=https%3A%2F%2Fen.wikipedia.org%2Fwiki%2FRotating_biological_contactor&docid=iw_qMbVP0hcHZM&w=373&h=206&hl=zh-CN&source=sh%2Fx%2Fim%2Fm5%2F4.

Approximately 95% of the surface area is thus alternately submerged in wastewater and then exposed to the atmosphere above the liquid. Carbon conversion may be completed in the first stage of a series of modules, with nitrification completed after the fifth stage. Most RBCs include a series of four or five modules to obtain nitrification. As the biofilm biomass changes from carbon metabolizing to nitrifying mode, its color changes from gray/beige to brown.

Biological growth is attached to the surface of the rotating disk and forms a slime layer. The rotating disks contact the wastewater with atmospheric air providing the oxidation mechanism. The rotation also helps to cast off excessive activated sludge. The disk system can be staged in series to obtain nearly any detention time or degree of removal required. Biofilms, which are biological growths that become attached to the disks, assimilate the organic materials (measured as BOD_5) in the wastewater. Aeration is provided by the rotating action, which exposes the media to the air after bringing it into contact with the wastewater, facilitating degradation of the pollutants being removed. The degree of wastewater treatment is related to the media surface area and the quality and volume of the inflowing wastewater.

Energy consumptions of the RBC and trickling filter are similar, both have relatively lower power than activated sludge with little noise owing to the slow rotation of 2–5 revolutions per minute. It is generally considered a very robust and low maintenance system. Its pollutant removal is likely lower than activated sludge or even trickling filter. This can be improved by adding a tertiary polishing filter after the RBC to lower BOD_5, suspended solids, and NH_3-nitrogen. Additional disinfection, such as an ultraviolet or chlorination process, can achieve effluent suitable for suitable reuse.

Constructed Wetland

A constructed wetland represents an alternative treatment system to conventional systems such as the activated sludge process. This is mainly used as tertiary treatment after the WWTP to further absorb nitrogen, phosphorus, and TSS. For agro-sectors covered in this study, it can be and has been used as a secondary treatment, e.g., for aquaculture effluent, given its low cost and ease of operation and maintenance (O&M), or as a tertiary treatment depending on wastewater characteristics.

All types of constructed wetland exhibit high treatment efficiency for organics and suspended solids comparable to conventional activated sludge. Removal of nitrogen depends on the type of constructed wetland and nitrogen species involved. The NH_3 is efficiently removed in vertical flow constructed wetlands while nitrate is removed efficiently in horizontal flow constructed wetlands. However, the combination of various types of constructed wetland (usually vertical and horizontal flow constructed wetlands) can enhance removal of total nitrogen and then the efficiency is comparable with conventional systems. Removal of phosphorus is variable depending on the filtration materials; however, commonly used materials do not support high phosphorus removal.

The constructed wetlands are often classified into two basic types: free water surface where the water surface is maintained at 10–50 centimeters (cm) above the constructed wetland bed; and subsurface flow where the water level is maintained below the constructed wetland bed. Common aquatic plants, such as cattail (*Typha* spp.), are used in constructed wetlands in Southeast Asia. Their desirable temperature is around 10–30°C and maximum salinity tolerance is 3% with optimum pH 4–10.

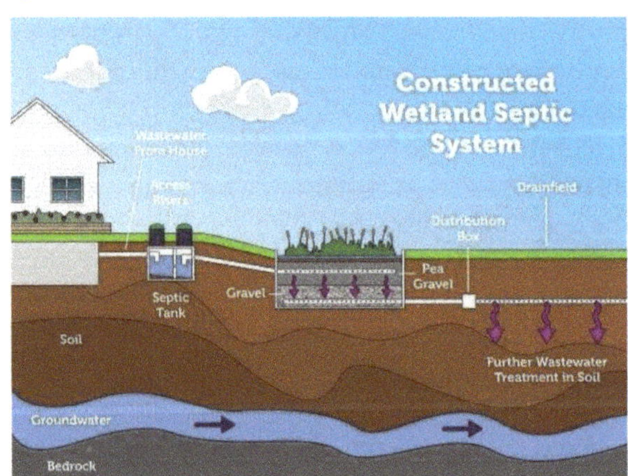

Figure A3: Septic Tank and Constructed Wetland

Source: United States Environmental Protection Agency. Septic Systems.

According to the US EPA, a free water surface constructed wetland typically includes one or more shallow basins or channels with a barrier to prevent seepage to sensitive groundwater and a submerged soil layer to support the roots of selected emergent vegetation, and inlet and outlet structures to distribute and collect wastewater, control water levels, and maintain hydraulic retention time. Wastewater at a relatively shallow depth of 10–50 cm flows over the vegetated soil surface, and the intended flow path through the system is horizontal.

Decentralized Wastewater Treatment System

All treatment technology described in this Appendix can be used as decentralized wastewater treatment, which utilizes mostly local materials and requires very little power and so is suitable for the small scale, typical for agro-sectors in rural areas. The decentralized wastewater treatment system (DEWATS) refers to technology developed by the German agency BORDA. It can serve for different users according to their discharge limits, from treatment processes of primary, secondary, to tertiary in various combinations of aerobic and anaerobic treatment processes. Simplicity is achieved through treatment without chemicals or energy using equipment with low O&M. Its maintenance activities can be carried out by service providers or by supervised and trained maintenance personnel.

DEWATS uses physical and biological treatment mechanisms such as sedimentation, flotation, and aerobic and anaerobic treatment to treat both domestic and industrial wastewater. Although it is claimed to be affordable, low maintenance with locally available materials, and without aeration thus low energy cost, there are some doubts about the actual performance of the DEWATS package provided in developing countries.

According to the providers, DEWATS can function as follows:

- primary treatment involving sedimentation and flotation;
- secondary anaerobic treatment in fixed-bed reactors including baffled upstream reactors or anaerobic filters;
- tertiary aerobic treatment in subsurface flow filters;
- tertiary aerobic treatment in polishing ponds;
- capacity to handle organic wastewater flows of 1–1,000 m^3/day; and
- systems are built to accommodate fluctuated loadings.

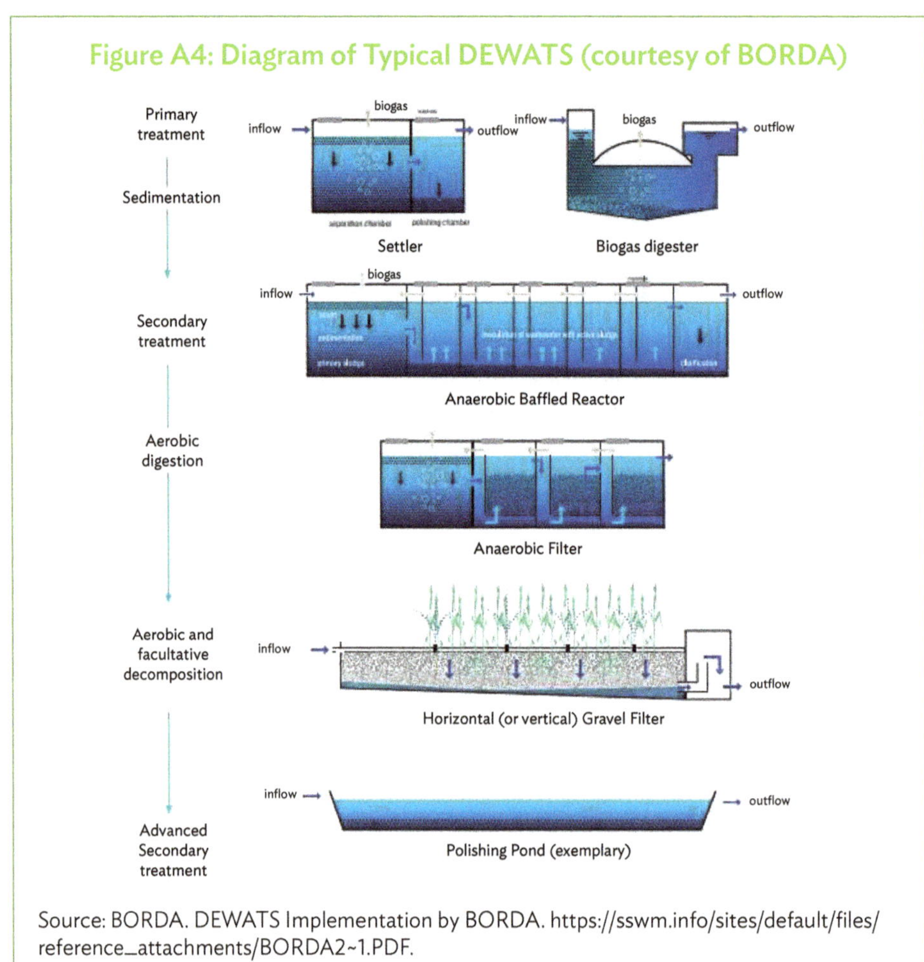

Figure A4: Diagram of Typical DEWATS (courtesy of BORDA)

Source: BORDA. DEWATS Implementation by BORDA. https://sswm.info/sites/default/files/reference_attachments/BORDA2~1.PDF.

Mini-Activated Sludge Module

Another decentralized wastewater treatment system has been promoted by the Japan Sanitation Consortium (JSC) under the name of Johkasou in Japanese. It is based on a septic tank but with aeration and settlement added—essentially a mini-activated sludge module, more suitable for households or small and medium-sized enterprises to meet stricter discharge limits. It basically converts the regular three-chamber septic tank into two anaerobic filter chambers and a third aerobic chamber with a blower providing aeration, and finally sedimentation and disinfection before the treated effluent flows out of outfall.

One of the key advantages of Johkasou is its compact size and low energy consumption, making it attractive for areas with limited space or resources. Johkasou can typically be made of reinforced concrete or fiber-reinforced plastic depending on their size and application. This can serve a single family to a community, with up to 100 m³/day capacity. Its innovative design allows downsizing the required tankage via adding a moving bed biofilm process, a biofilm filtration process, and a moving bed biofilm filtration process to replace the traditional contact aeration process.

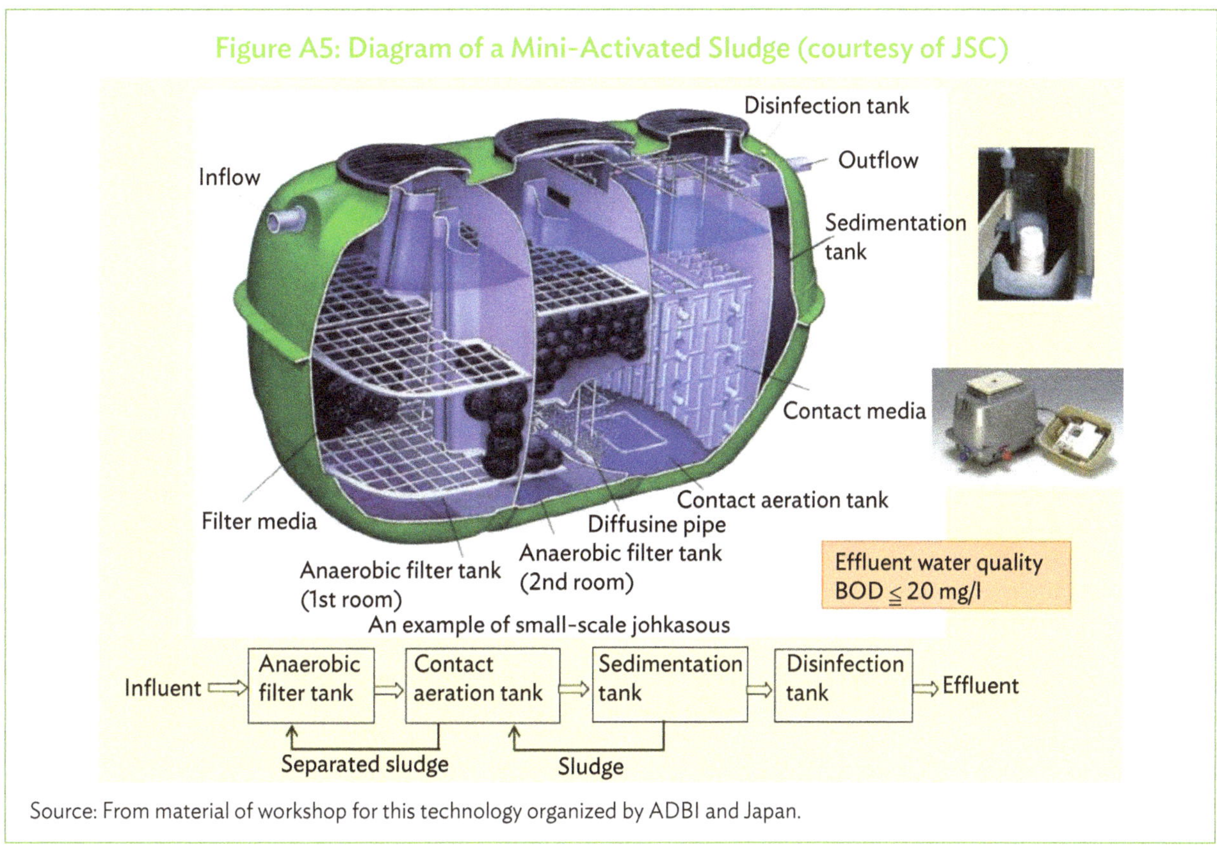

Source: From material of workshop for this technology organized by ADBI and Japan.

Aerated Lagoon

A lagoon is a pond of engineering design typically made of earth without metal or concrete tanks and thus low in CAPEX. The lagoon system O&M is minimal and flow through the system is usually by gravity, unless recirculation is required. Recirculation can reduce the buildup of bottom solids near the inlet of the influent unit. Bar screens are usually installed at its influent unit and it is also equipped with an easy flow measurement instrument, e.g., a Parshall flume, to monitor the flows entering the lagoon.

The upper layer of the lagoon is usually facultatively aerobic whereas anaerobic conditions exist near the bottom. Facultative microorganisms can function under either aerobic or anaerobic conditions. A series of lagoons is frequently used, and their number and size are according to final effluent requirements, incoming waste load, temperature, and climate conditions.

Two anaerobic ponds produce a higher quality effluent than a single pond, thus reducing the total loading and size of the facultative pond. An anaerobic pond is sometimes operated for half a year as the anaerobic digester. The lagoon system can generally have additional lagoons in series for additional treatment.

Mechanically aerating via surface floating aerators for the oxidation pond improves treatment and reduces the pond size, hence being named an aerated lagoon. This is followed by a maturation pond, which basically performs a similar function to the secondary sedimentation tanks of the standard activated sludge. The series of flows in ponds can create a buffer against shock loadings, which is one of the advantages of the pond treatment process. In an aerated lagoon system, the aerators provide two functions: transferring air into the basins required by the biological oxidation reactions, and the mixing required for dispersing the air and for contacting the reactants (i.e., oxygen, wastewater, and microbes). However, they do not provide as good mixing as normally achieved in activated sludge systems.

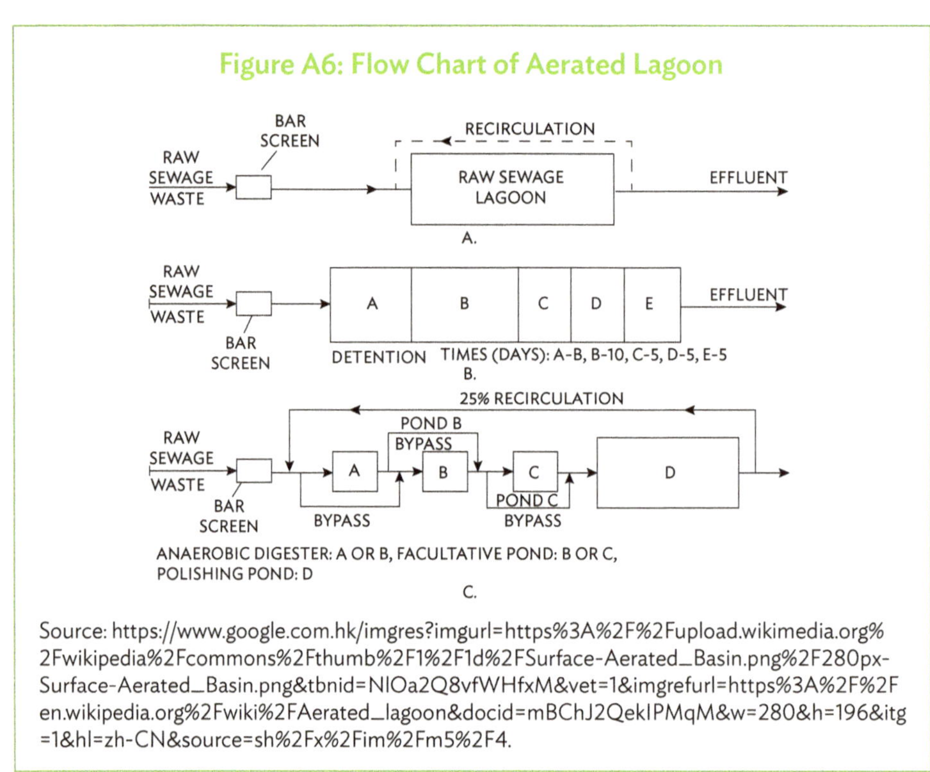

Figure A6: Flow Chart of Aerated Lagoon

Source: https://www.google.com.hk/imgres?imgurl=https%3A%2F%2Fupload.wikimedia.org%2Fwikipedia%2Fcommons%2Fthumb%2F1%2F1d%2FSurface-Aerated_Basin.png%2F280px-Surface-Aerated_Basin.png&tbnid=NlOa2Q8vfWHfxM&vet=1&imgrefurl=https%3A%2F%2Fen.wikipedia.org%2Fwiki%2FAerated_lagoon&docid=mBChJ2QekIPMqM&w=280&h=196&itg=1&hl=zh-CN&source=sh%2Fx%2Fim%2Fm5%2F4.

Advantages and disadvantages of aerated lagoons are listed below:

- cost-effective option in areas where land is inexpensive;
- use less energy than most wastewater treatment methods;
- simple O&M and generally requires only part-time staff;
- can handle intermittent use and absorb shock loadings;
- very effective to remove disease-causing organisms (pathogens);
- effluent is suitable for irrigation (where appropriate) owing to its high nutrient content;
- requires more land than other wastewater treatment methods;
- less efficient in cold climates and require longer detention times in these areas;
- attracts mosquitoes and other insects if not adequately maintained;
- odor arises during algae blooms or if anaerobic lagoons inadequately maintained; and
- effluent from certain lagoons contains algae and may require additional treatment or "polishing" to meet local discharge standards.

Oxidation Ponds

Also known as lagoons or waste stabilization ponds, oxidation ponds are shallow and large ponds normally without any aeration but utilize sunshine, microorganisms, and algae to create oxygen needed for bio-decomposition of organic matter. The microorganisms present in the oxidation pond oxidize organic matter and result in the release of carbon dioxide (CO_2), water, and NH_3, etc. The algal development occurs when sunshine is present, and utilizes the inorganic waste produced by the breakdown of organic matter and releases oxygen. Microorganisms consume the biodegradable organics using the oxygen generated by the algae, and CO_2 is generated at the same time. Eventually, algae use CO_2 to reduce inorganic waste such as nitrogen and phosphorus. As oxidation and reduction reactions occur simultaneously, an oxidation pond is also referred to as a "redox pond."

The remaining non-degradable or solid organic wastes settle as sludge to the bottom of the ponds and then convert insoluble organic waste into soluble organic acids, such as ethanol by anaerobic microorganisms. Finally, anaerobic microorganisms decompose organic acids further and release hydrogen sulfide, methane, NH_3, and CO_2 etc. In addition, excessive sludge from wastewater treatment in the oxidation ponds can be used as fertilizer.

For the oxidation of biodegradable organic waste, oxygen supply must be sufficient. The heterotrophic bacteria satisfy their oxygen needs with the oxygen supplied by algae and the oxygen in the atmosphere. That is why mechanical aerators are occasionally installed to provide more oxygen and so reduce the required pond size, effectively converting it to an aerated lagoon/pond. Eventually, excessive sludge needs to be removed through dredging or pumping. Remaining algae in the pond effluent can be eliminated using chemical treatment, settling, filtration, disinfection, or the combination of these processes.

The oxidation pond is straightforward and is the cheapest secondary treatment process and does not require complicated equipment. In tropical regions, oxidation ponds are a viable and efficient way to treat wastewater. The routine O&M does not require a great deal of labor.

The disadvantage is the requirement of extra land space, which should be well considered within the overall investment. In addition, the potential foul odor from the ponds in different seasons also needs to be well controlled. The site selected needs to have sufficient clayish soil with low permeability. The earth of the site requires sufficient compacting or other treatment to further reduce permeability to minimize potential contamination over the years.

Development of Environmental Standards in the United States

The United States Environmental Protection Agency (US EPA) is responsible for overall management of the environment in the US. In consideration of the geographic situations and various factors of natural conditions, the US EPA set up 10 regions over the 50 states and autonomous territories in dealing with their respective and associated environmental issues. Certain states and territories were chosen to be in the same region, primarily based on location and climate zone.

Each region develops their own environmental standards to meet all requirements in relation to the environmental issues, specifically for the region. For example, a wastewater treatment plant (WWTP) requires its own discharge permit (National Pollutant Discharge Elimination Permit [NPDES]) not just based on a set of national limits but also on its location and surrounding environment. Sometimes, a river basin management plan or environmental master plan is the governing document for any level of government agencies to follow and to determine the required permit of a WWTP. There is no fixed discharge limits for a WWTP, but it must achieve its operational outputs, mostly in accordance with the location and adjacent waters to receive its treated effluent.

Wastewater treatment development in the US went through an early stage of primary treatment process initially based on the situation of environmental protection objectives and associated discharge limit revisions. This was followed by advanced primary treatment after limits were tightened, then secondary (biological) treatment and eventually tertiary treatment processes were introduced when requirements became more stringent. Although treatment processes and technologies become more capable in dealing with the stringent discharge limits, the permit for each WWTP is not set up as one single set of numbers but depends on its service area and where the effluent will be discharged to, i.e., the adjacent waters. For WWTPs without suitable waters to discharge most likely its permit will require zero discharge. However, when receiving water has high assimilation capacity, its permit can be less stringent.

At the state and county levels, the Regional Water Quality Control Board (RWQCB) is typically the agency responsible for overseeing the discharge limits of any WWTPs and ensuring relevant permits are in compliance. Normally, all WWTPs submit their monthly or quarterly effluent reports to the RWQCB and report any noncompliance with the possible scenarios and/or methods with a time-bound action plan to resolve the issues. When the committed time arrives, the subject WWTP submits their effluent monitoring report with adequate proof provided.

For example, the Point Loma WWTP located in San Diego, California mostly treats domestic wastewater with a capacity of almost 1 Mt per day. Its NPDES permit requests that, for both biochemical oxygen demand and total suspended solids, the 30-day average shall not exceed 30 mg/L and removal shall not be less than 85%. These are not standards from any federal or state regulations but are based on annual effluent concentrations in the previous permits.

This perfectly represents the concept of a specific permit for individual WWTPs and case-by-case in accordance with its service area and where its effluent will be discharged to. Most importantly, it will be sustainable according to its own performance in line with its specific permit.

If a WWTP does not comply with its discharge permit, it is given a reasonable time to resolve the issue, or a citation is given if the noncompliance issues still exist. In addition, occasionally the RWQCB will visit WWTPs without notice and take samples of effluent to check on plant performance, to ensure all required discharge limits are in compliance with the permit. This un-notified visit plays a critical role not only to validate the WWTP performance but also to alert all WWTPs to monitor their routine operation and maintenance activities on a daily basis.

In summary, there is no one set of discharge limits for all WWTPs but a permit for each individual WWTP in accordance with their service area and the environment requirements of the receiving waters where treated effluent is discharged to. With this approach, each WWTP must ensure that all their facilities—sewer collection, pumping (booster and influent), and plant operation and maintenance—function as one system.

References

Bachi, O. E. et al. 2022. Wastewater Treatment Performance of Aerated Lagoons, Activated Sludge and Constructed Wetlands under an Arid Algerian Climate *Sustainability*. Multidiscipline Digital Publishing Institute.

Bahia, Y. et al. 2019. Estimation of Greenhouse Gas (GHG) Emissions from Natural Lagoon Wastewater Treatment Plant: Case of Ain Taoujdate Morocco. *Environment, Energy and Earth Sciences (E3S) Web of Conferences*. 150 (010).

Bosak, V. et al. 2016. Performance of a Constructed Wetland and Pretreatment System Receiving Potato Farm Wash Water. *Water*. Multidiscipline Digital Publishing Institute.

Elkheir, O. et al. 2022. Wastewater Treatment Performance of Aerated Lagoons, Activated Sludge and Constructed Wetlands under an Arid Algerian Climate. *Sustainability*. Multidiscipline Digital Publishing Institute.

Fang, S. et al. 2004. Technology of Aquaculture Wastewater Treatment and Application. Techniques and Equipment for Environmental Pollution Control.

Government of the United Kingdom. Wastewater Discharge Standard. Waste Water Treatment Works: Treatment Monitoring and Compliance Limits. www.gov.uk.

Hassard, F. et al. 2015. Rotating Biological Contactors for Wastewater Treatment - A Review. *Process Safety and Environmental Protection*. 94.

IFC. 2008. Environmental, Health and Safety Guidelines for Health Care Facilities. Washington, DC.

Kollu, V. K. R. et al. 2022. Comparison of Microwave and Autoclave Treatment for Biomedical Waste Disinfection. Systems Microbiology and *Biomanufacturing*. 2. pp. 732–742.

Kurniawan, S. B. et al. 2021. Aquaculture in Malaysia: Water-Related Environmental Challenges and Opportunities for Cleaner Production. *Environmental Technology & Innovation*. 24.

Lopez-Sanchez, A. et al. 2022. Microalgae-based Livestock Wastewater Treatment as a Circular Bioeconomy Approach, Enhancement of Biomass Productivity, Pollutant Removal and High-Value Compound Production. *Journal of Environmental Management*. 308.

Metcalf and Eddy. 2004. Wastewater Engineering: Treatment and Reuse. 4th Edition. New York: McGraw Hill.

Miller, R. et al. 2011. How a Lagoon Works For Livestock Wastewater Treatment. *Agriculture*. Utah State University Cooperative Extension.

Mozhiarasi, V. et al. 2022. Slaughterhouse and Poultry Wastes: Management Practices, Feedstocks for Renewable Energy Production, and Recovery of Value Added Products. *Biomass Conversion and Biorefinery*.

Parihar, S.S. et al. 2019. Livestock Waste Management: A Review. *Journal of Entomology and Zoology Studies*. 7 (3). pp. 384–393.

Parzentna-Gabor, A. 2021. Odor Removal Technologies. Scholarly Community Encyclopedia.

Pryce, D. et al. 2022. Economic Evaluation of a Small Wastewater Treatment Plant Under Different Design and Operation Scenarios by Life Cycle Costing. *Development Engineering*. 100103.

Puchlik, M. 2016. Application of Constructed Wetlands for Treatment of Wastewater from Fruit and Vegetable Industry. *Journal of Ecological Engineering*. 17 (1).

Ren, X. 2022. Supporting One Health through Environmental Safeguards. *Development Asia*. https://development.asia/insight/supporting-one-health-through-environmental-safeguards.

Reyes, I. P. et al. 2015. Anaerobic Biodegradation of Solid Substrates from Agroindustrial Activities - Slaughterhouse Wastes and Agrowastes. INTECH.

Sekandari, A. W. 2019. Cost Comparison Analysis of Wastewater Treatment Plants. *International Journal of Science Technology & Engineering*. 6 (1).

Senatore, V. et al. 2021. Full-Scale Odor Abatement Technologies in Wastewater Treatment Plants (WWTPs): A Review. *Water*. 13 (24). p. 3503.

Sheng, W. et al. 2022. Review on Treatment Technology of Aquaculture Wastewater and Its Prospect. *Fishery Information and Strategy*. 31 (1).

Su, G. et al. 2021. Valorisation of Medical Waste Through Pyrolysis for a Cleaner Environment: Progress and Challenges. *Environmental Pollution*. 279. 116934.

Tay, J.-H. et al. 2004. Seafood Processing Wastewater Treatment. In L. K. Wang et al., eds. *Handbook of Industrial and Hazardous Wastes Treatment*. Cleveland, Ohio: Cleveland State University.

Tom, A. P. et al. 2021. Aquaculture Wastewater Treatment Technologies and Their Sustainability: A Review. *Energy Nexus*. 4. 100022.

Tony, M. A. et al. 2011. The Use of Solar Energy in a Low-Cost Drying System for Solid Waste Management: Concept, Design and Performance Analysis.

United States Environmental Protection Agency (US EPA). 2000. National Effluent Guidelines. Code of Federal Regulations. Title 40, Part 437 - The Centralized Waste Treatment Point Source Category.

Kollu, V. K. R. et al. 2022. Comparison of Microwave and Autoclave Treatment for Biomedical Waste Disinfection. Systems *Microbiology and Biomanufacturing*. 2. pp. 732–742.

Vanotti, M. B. et al. 2008. Greenhouse Gas Emission Reduction and Environmental Quality Improvement from Implementation of Aerobic Waste Treatment Systems in Swine Farms. 28 (4). pp. 759–766.

Water Environment Federation (WEF). 2009. Manual of Practice 8. Design of Municipal Wastewater Treatment Plants. Fifth edition. McGraw-Hill Professional.

Waqas, S. et al. 2023. A Review of Rotating Biological Contactors for Wastewater Treatment. Water. Multidiscipline Digital Publishing Institute.

Xiong, R. et al. 2019. Rural Domestic Sewage Treatment by a Combined Process of Anaerobic Tank, Drop-Aeration and Constructed Wetland. *Chinese Journal of Environmental Engineering*. 13 (2). pp. 327–331.

Zhang, Y. et al. 2015. Performance of System Consisting of Vertical Flow Trickling Filter and Horizontal Flow Multi-Soil-Layering Reactor for Treatment of Rural Wastewater. *Bioresource Technology*. 193. pp. 424–432.